The Rangeley and Its Region

The Rangeley
and Its Region

The Famous Boat and Lakes of Western Maine

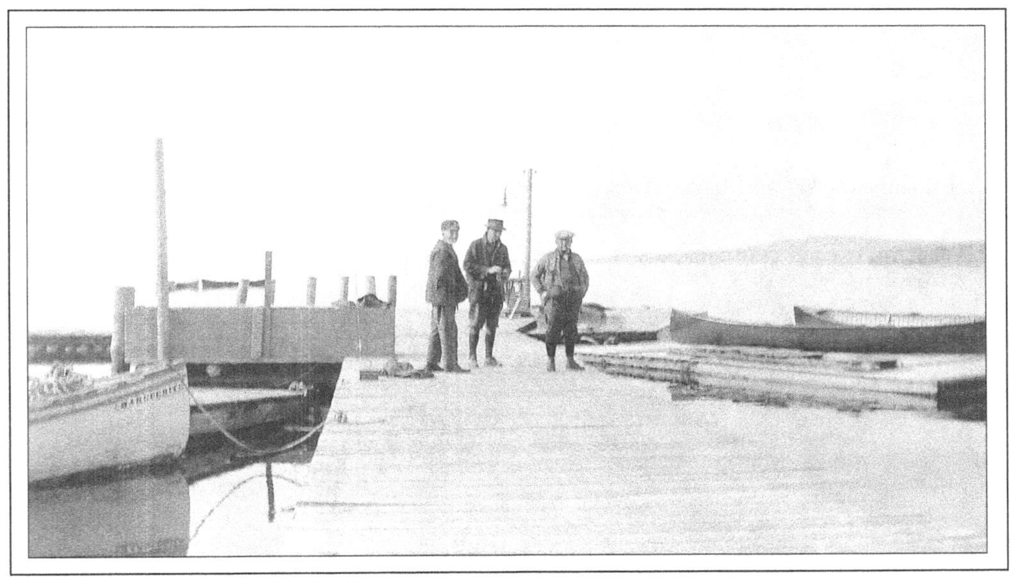

Stephen A. Cole

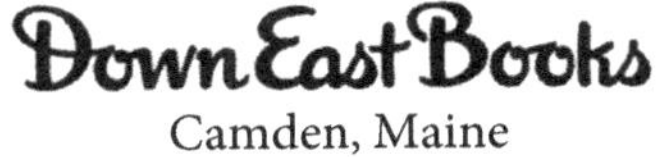

Camden, Maine

Down East Books

An imprint of Globe Pequot

Distributed by NATIONAL BOOK NETWORK

First paperback edition: 2007
First Down East Books edition: 2023

Library of Congress Cataloging-in-Publication Data

Cole, Stephen A., 1955-

  The Rangeley and its region : the famous boat and lakes of western Maine / Stephen A. Cole.
      p. cm.
  Includes bibliographical references.
  ISBN 978-1-68475-168-6 (pbk.)

  1. Rangeley boats—History. 2. Brook trout fishing—Maine—Raingeley Region—Equipment and supplies—History. 3. Brook trout fishing—Rangeley Lakes (Me. and N.H.)—Equipment and sup-plies—History. 4. Rangeley Region (Maine)—History. 5. Rangeley Lakes Region (Me. and N.H.)—History. I. Title.

  SH689.3.C65 2007
  799.1'10974172—
  dc22

Front cover photograph: Rangeleys at the Upper Dam Pool *(Rangeley Historical Society)*
Back Cover photographs: A sport, a guide, and Rangeley boats *(Rangeley Historical Society)*
Map of the Rangeley region *(Rangeley Historical Society)*
Title page photograph: Haines Landing *(Rangeley Historical Society)*

Design and layout by Lindy Gifford, Damariscotta, Maine
Copyediting by Genie Dailey, Fine Points Editorial Services, Jefferson, Maine

Contents

Introduction

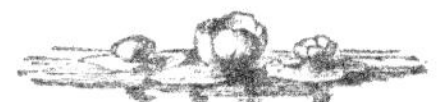

On side tables, desks, and fireplace mantels in the town of Rangeley, Maine, sit small brass castings of a long and narrow open boat. The boat is called a Rangeley. In 1976, when the country was celebrating its two-hundredth birthday, Rangeley citizens met to decide what memento they might commission which would be symbolic of their home and its history. They chose the distinctive double-ended boat that was the namesake of the region, the boat that had propelled sportfishing on the Rangeley Lakes for one hundred years. Indeed, the Rangeley boat is the one true icon of this great semi-wilderness.

It was angling that brought Rangeley, Maine, to prominence in the nineteenth century, and brought the Rangeley boat into being. Sportfishing remains a significant part of Rangeley's life and economy. Here is a town in which books such as *McClane's Standard Fishing Encyclopedia, Streamer Fly Tying and Fishing,* and *Trout Fishing in America* all reside in the reference section of the public library, while *Fly Fisherman* and *Trout* magazines are found among the periodicals. At the hardware store in downtown Rangeley, one can choose from among seven different styles of nets for landing trout and salmon, and even the smallest variety store stocks streamer flies and plastic boxes bristling with dozens of wet and dry flies, tied locally. Prominently displayed at the front counters of these same stores are small photo albums placed there by local fishing guides. Inside are color snapshots of sizable brook trout and landlocked salmon held aloft by happy anglers as testament to the guides' talents. The Rangeley boat has always been an intimate part of this fishing culture, a boat designed for no other purpose than fishing these famous lakes. What follows is not so much a record of how the Rangeley boat was built—for that is known—but of how it evolved, who built it, how it was used, and its place in the life of the region.

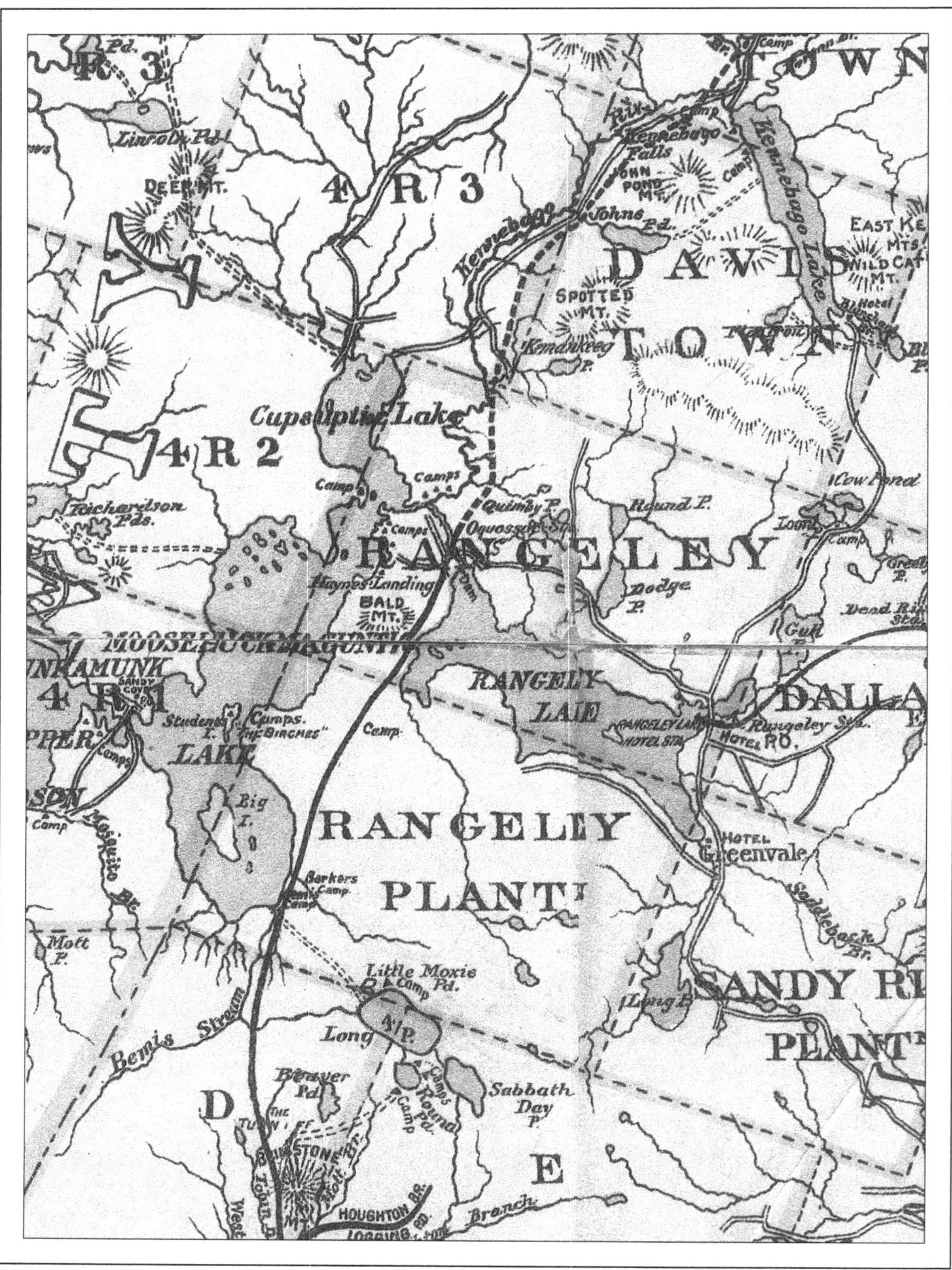

The Rangeley Lakes *(Rangeley Historical Society)*

Beginnings: Rangeley and the Oquossoc Angling Association

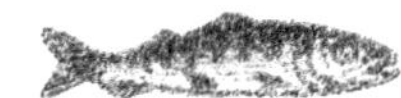

Rangeley's origins were no different than those of many remote northern New England towns. Settlers trickled into the region during the early nineteenth century, establishing themselves on lands purchased by James Rangeley, Sr., of Philadelphia and several partners from the Commonwealth of Massachusetts in 1796. What the first arrivals found were vast conifer woodlands set amid the Appalachian Mountains and dominated by five great lakes. The state of New Hampshire was close by to the west, and at the north was Canada. Water was everywhere; within a ten-mile radius of the Rangeley settlement were forty lakes and ponds. But it was the big lakes, linked by streams, that commanded attention. These headwaters of the Androscoggin River covered more than sixty-four square miles and seemed more like inland seas than mere lakes. They were known by aboriginal names—Oquossoc, Mooselookmeguntic, Cupsuptic, Mollychunkamunk, and Welokennebacook—but with time and the pervasion of American culture, Oquossoc would become Rangeley Lake and Mollychunkamunk and Welokennebacook, the Richardsons.

Rangeley's first economy was ordinary, based on lumbering and agriculture. It was providential that other means of livelihood soon came along, as the region's growing season was a scant ninety days. Winters seemed unending, the temperature sometimes fell to twenty below zero, and the lakes froze to a depth of three feet. The ice didn't go out until the first or second week of May, and sometimes later.

Early visitors to the lakes generally came with fishing rods. Gentlemen from Providence, Rhode Island, and New York fished Rangeley and Mooselookmeguntic lakes during the 1840s, and in the following decade Yale collegians exploring these waters put up a camp on an island in Mooselookmeguntic which remains known as Students Island. The catalyst for change in Rangeley was a visit paid to the lakes in 1860 by Henry

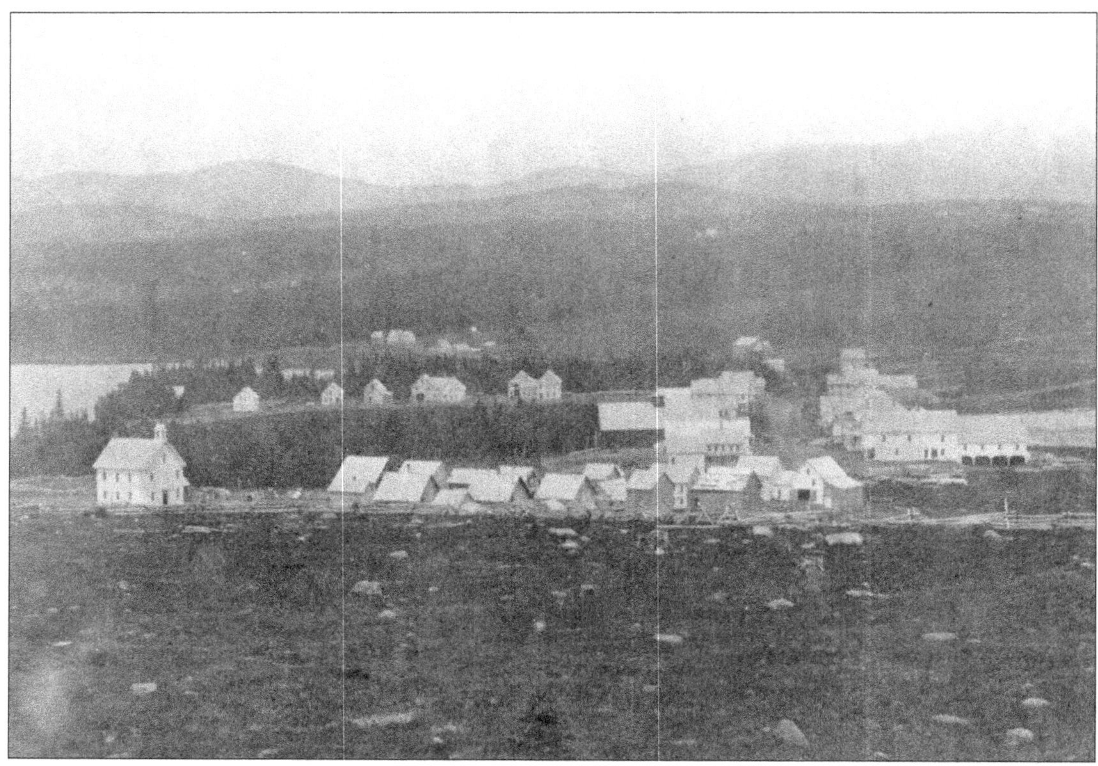

Nineteenth–century Rangeley *(Maine Historic Preservation Commission)*

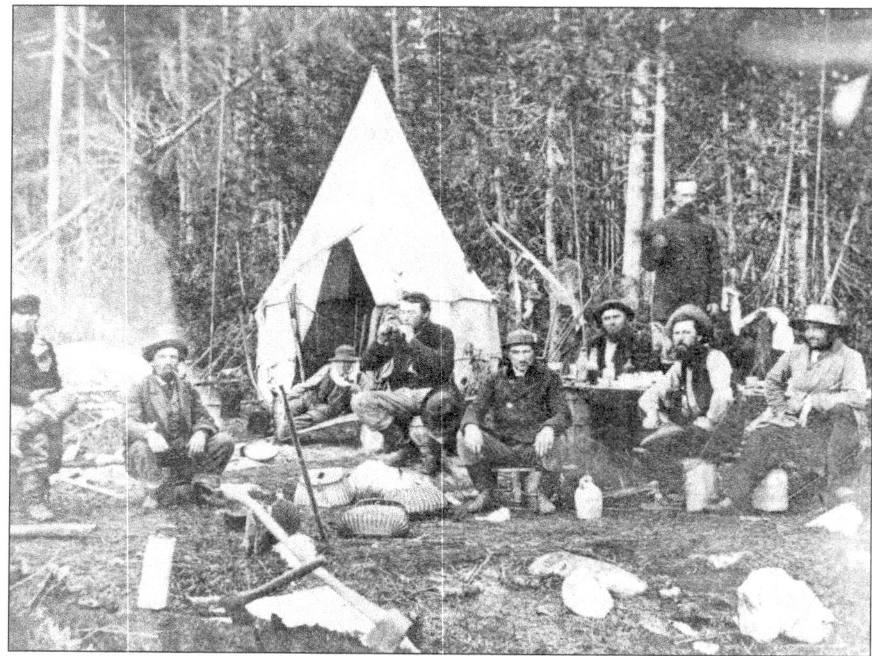

"Camping Out" at Upper Dam, Upper Richardson Lake, circa 1870–80 *(Maine Historic Preservation Commission)*

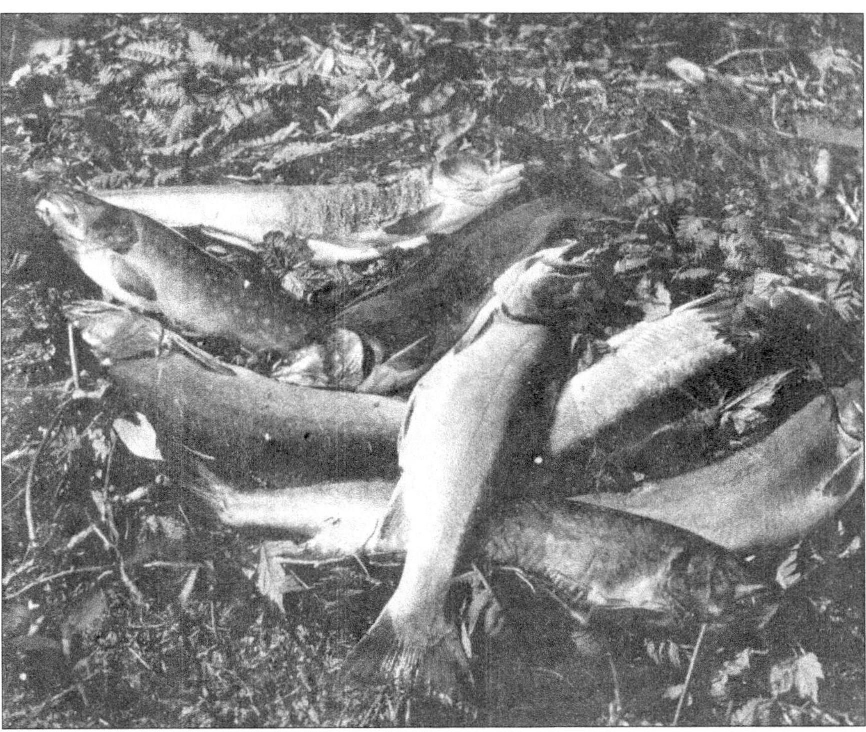

Brook trout, first day's catch *(Maine Historic Preservation Commission)*

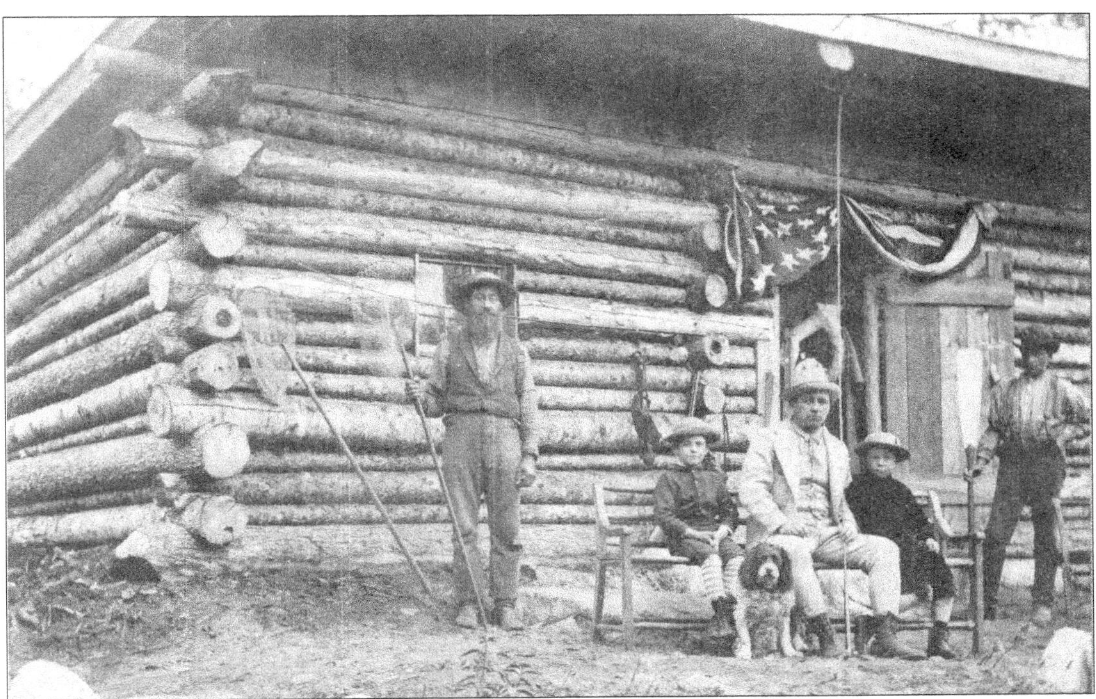

New Yorker George Shepard Page and young at his camp, Bemis Stream
(Maine Historic Preservation Commission)

Stanley (who would later become Maine's commissioner of fisheries) and George S. Page, a New York sportsman. Stanley's father had been one of those who had traveled to Rangeley a generation earlier and been astonished at the size and abundance of brook trout; his son's trip was no different. What they observed were lakes that had never really been fished, lakes that provided optimum conditions for producing large fish. George Shepard Page did not forget the extraordinary trout he'd caught on the big lakes of western Maine, and he returned several years later. When he left for home that time, he carried with him eight brook trout packed in ice and birch bark and sawdust—the smallest weighing five pounds, the greatest more than eight pounds.

Back in New York City, Page presented the largest trout to a small group of notable newspapermen, including William Cullen Bryant of the *Evening Post* and Henry J. Raymond of the *New York Times*. As Page must have expected, news of the remarkable catches found its way into print and caused both a sensation and controversy. The furor surrounded the sheer size of these fish. The best brook trout fishing of the day was to be had in the Adirondacks, where the finest fish never surpassed five pounds; many sportsmen simply concluded that George Page had mistaken the larger lake trout for brook trout. But George Page's vindication came from the fish he had shipped to the renowned biologist Louis Agassiz at Harvard. Agassiz examined the specimen and pronounced it none other than *salvelinus fontinalis,* the brook trout. From then on, the Rangeley Lakes would be another of the summer places that wealthy, urbane Victorians were fashioning out of the wilderness, a mecca for eastern anglers hoping to spar with record trout in a pristine place.

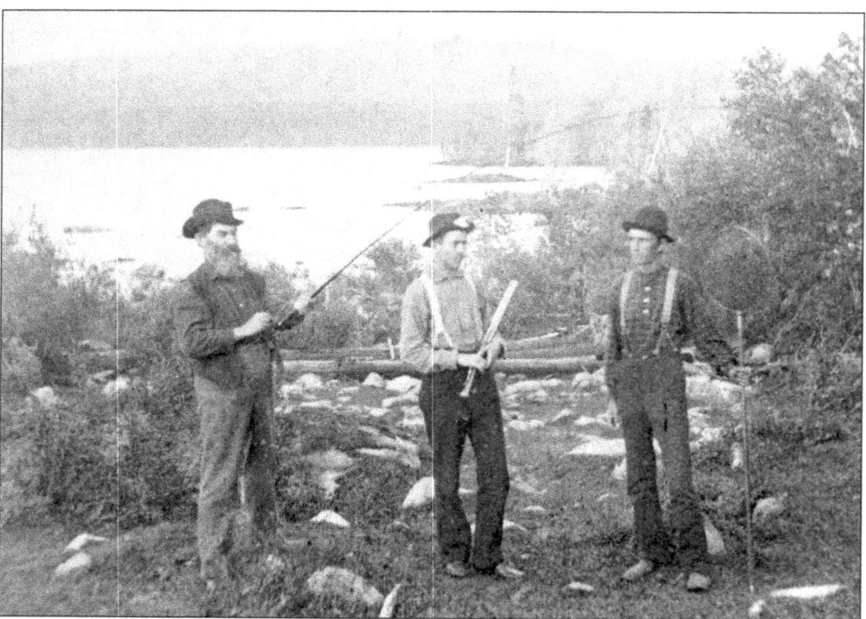

Early anglers, circa 1870–80 *(Maine Historic Preservation Commission)*

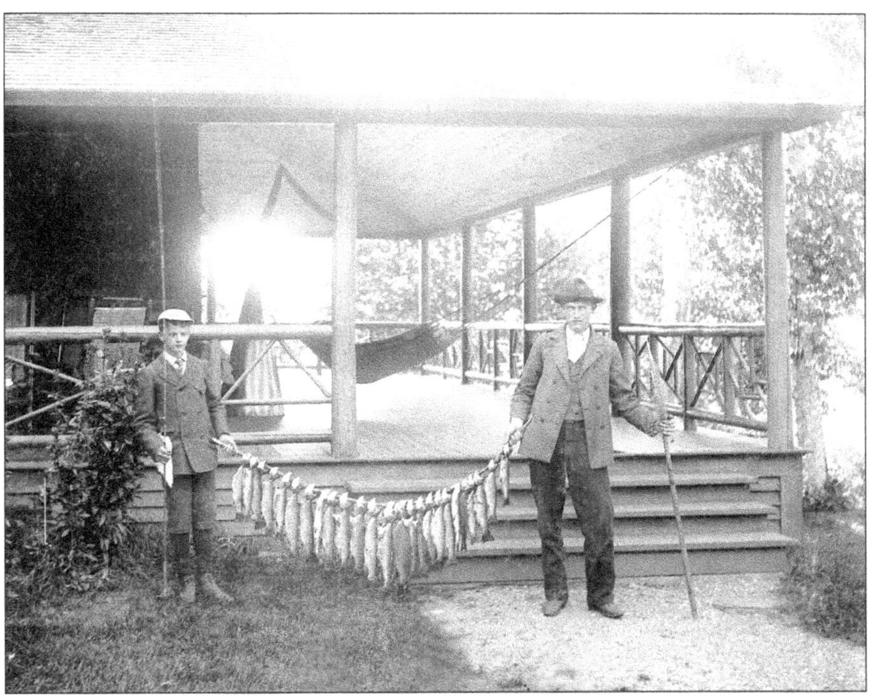

Man, boy, and brook trout *(Rangeley Historical Society)*

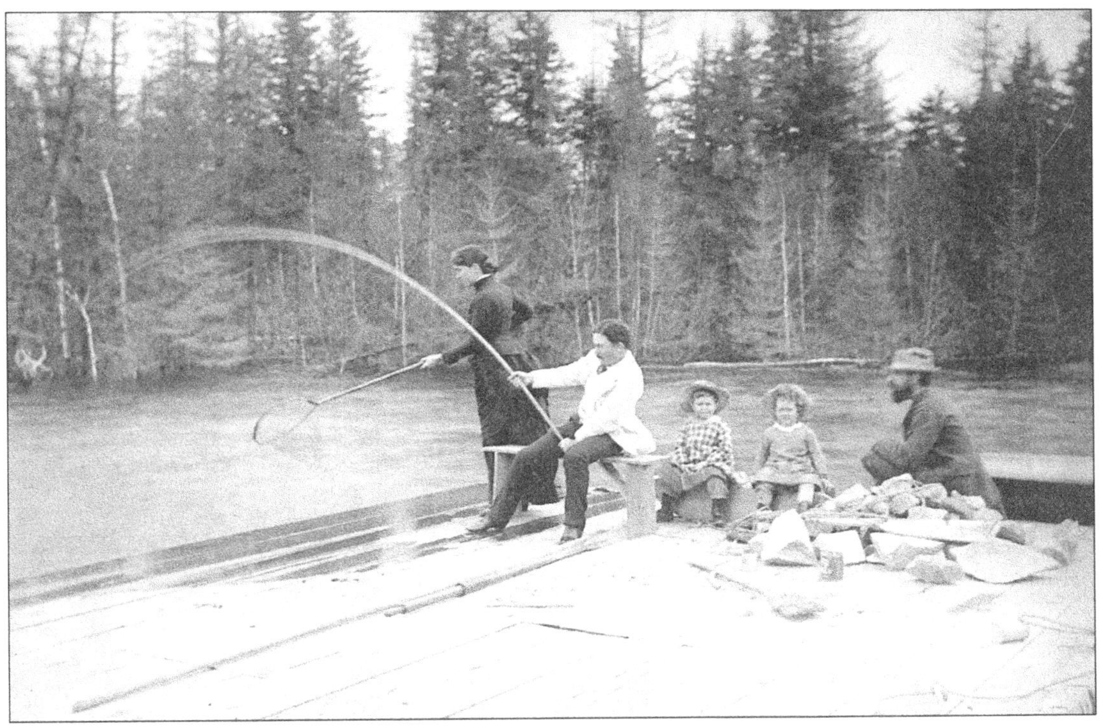

Rustic entertainment *(Rangeley Historical Society)*

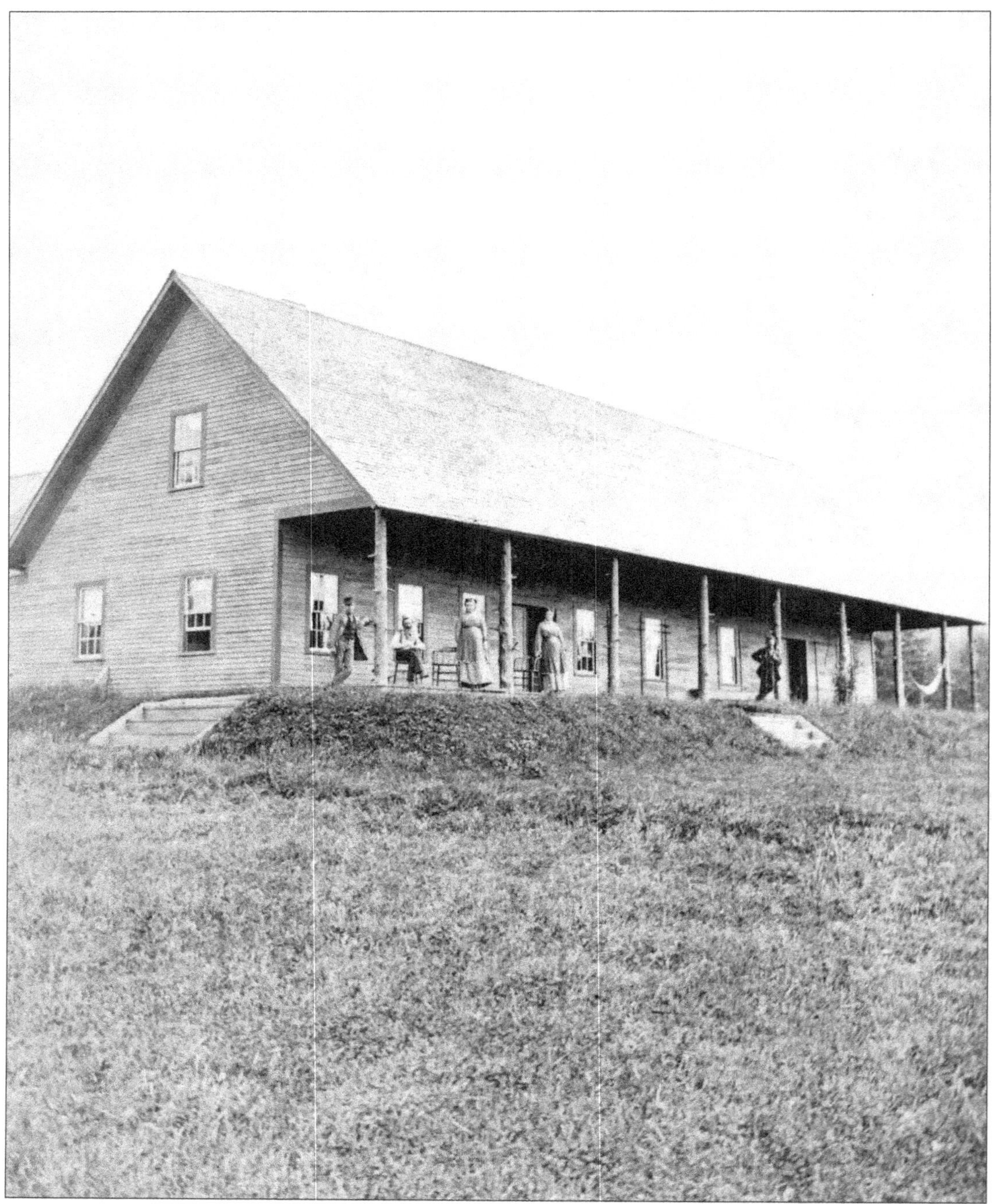

Camp Kennebago, Oquossoc Angling Association, circa 1873–78
(Maine Historic Preservation Commission)

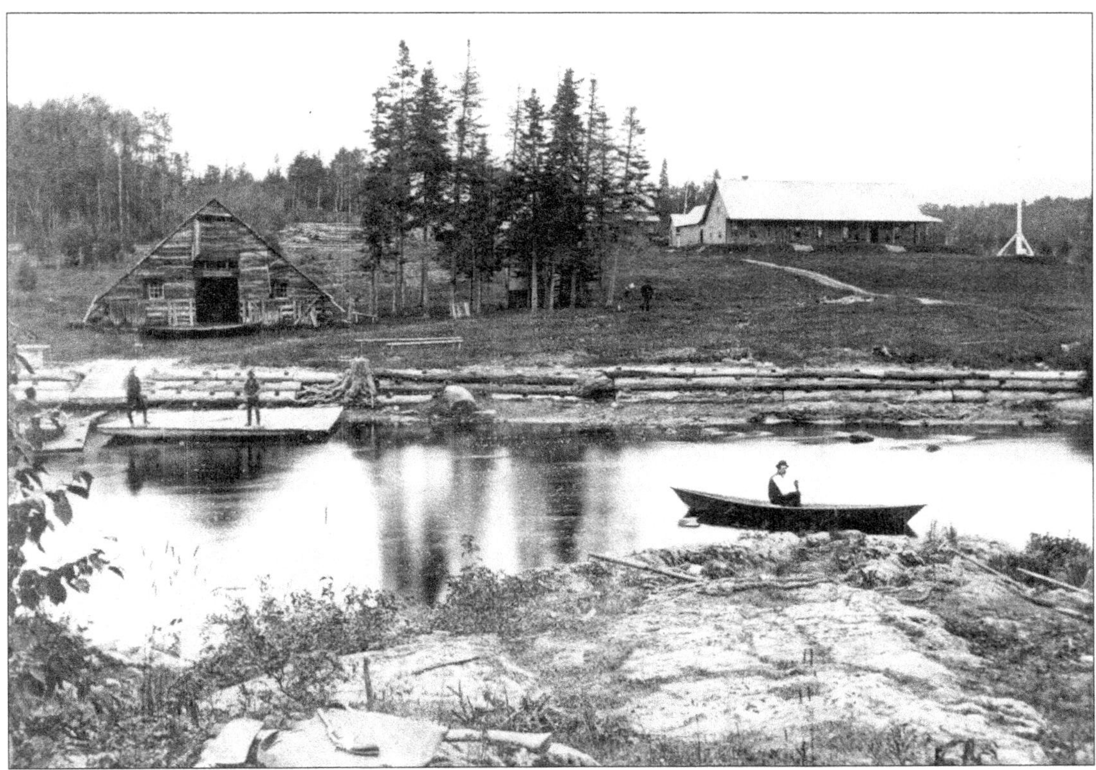

Indian Rock, Camp Kennebago, and bateau or Umbagog boat, circa 1870–80,
(Maine Historic Preservation Commission)

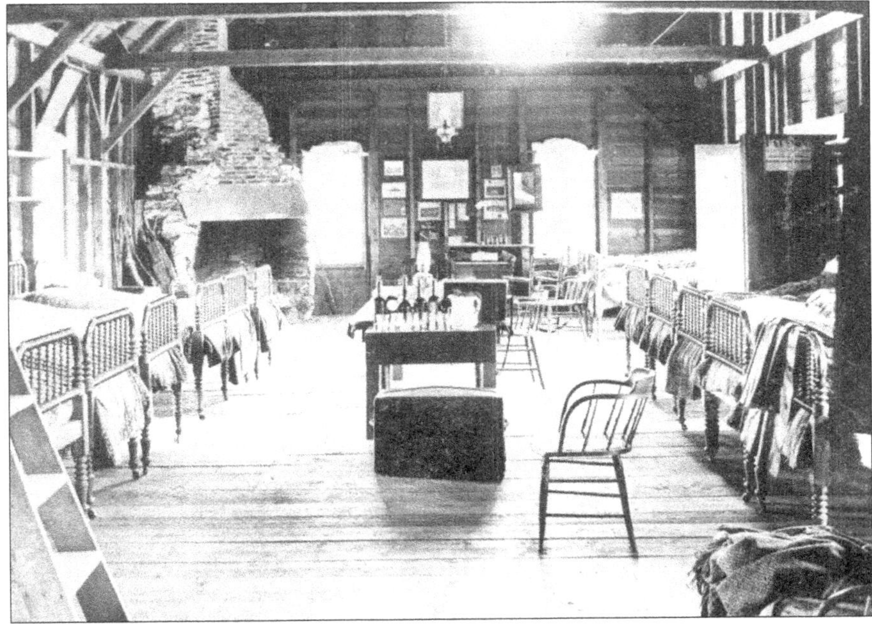

Camp Kennebago interior, circa 1873–78
(Maine Historic Preservation Commission)

As important to sportfishing as George Page's newsworthy trout were the beginnings of a small fishing club in which Mr. Page also had a hand. In 1868 he joined with others from New York and New England and a gentleman from New Orleans in founding the Oquossoc Angling Association, whose goal was to enjoy and preserve the splendid fishing of the local waters. Within a year the O.A.A. had acquired considerable acreage encompassing trout ponds and streams and a defining place: land at the confluence of the Rangeley and Kennebago rivers, about seven miles west of Rangeley Village near the hamlet called Oquossoc. Close to the shore they constructed Camp Kennebago, a large, plain building that served the fishermen as dining room, dormitory, and clubhouse. Later would come more buildings, a camp for the guides, a dormitory for chauffeurs and housemaids, the boathouse, and, finally, the private camps built by members and their families. Painted a summery yellow, but obscured by unruly bushes at its front, Camp Kennebago still stands at the juncture of the rivers, a proper symbol of America's oldest fishing club. Soon after its founding, the Oquossoc Angling Association did two things of lasting importance to the Rangeley Lakes: It introduced landlocked salmon to the waters and went about the business of finding a suitable boat for fishing there.

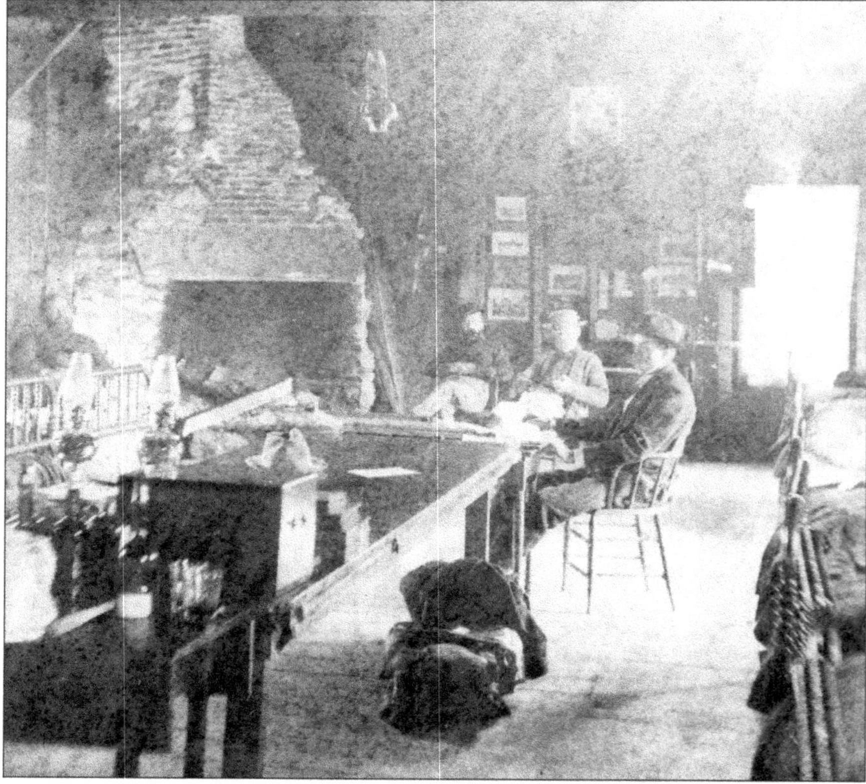

Members relaxing, Camp Kennebago, 1870–80
(Maine Historic Preservation Commission)

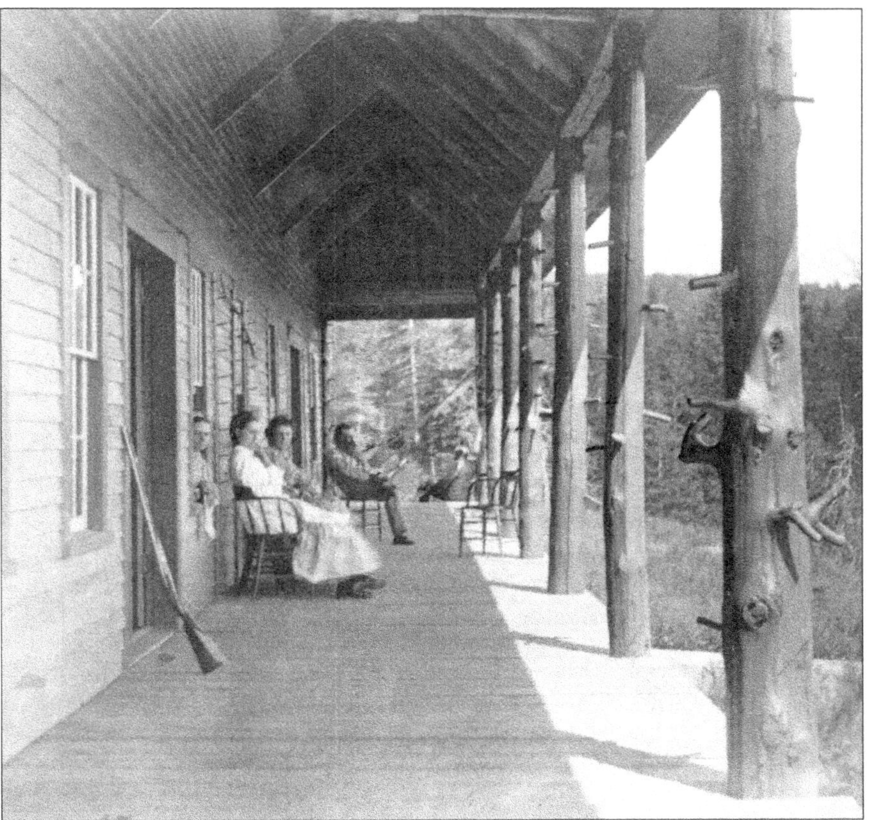

On the piazza, Camp Kennebago *(Maine Historic Preservation Commission)*

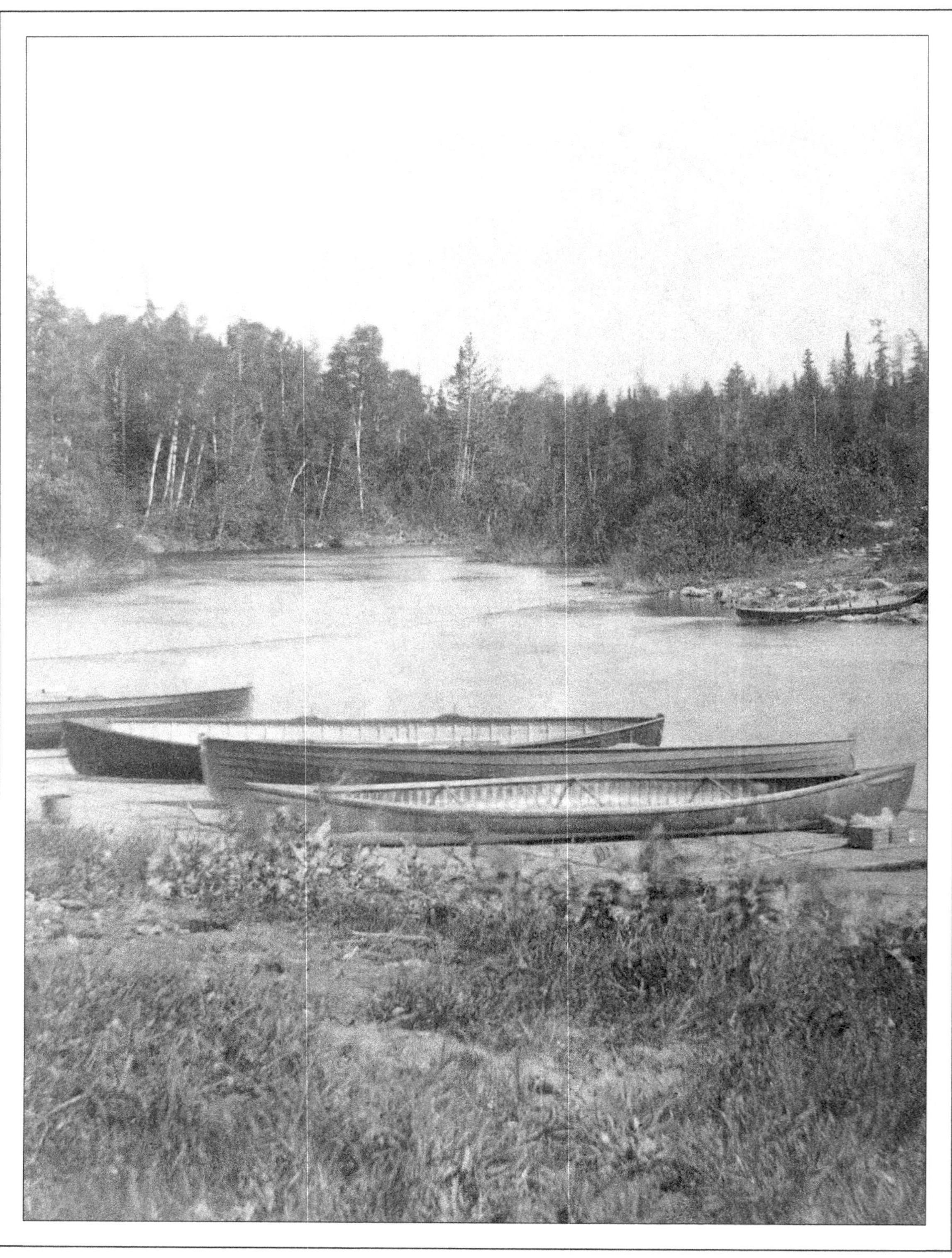

"Indian Rock" boats and a canoe on the dock, Oquossoc Angling Association, junction of the Kennebago River and Oquossoc Stream *(Maine Historic Preservation Commission)*

The Rangeley Boat

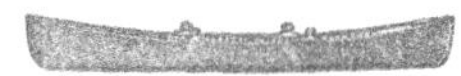

Making boats, like making humans, governments, towns, and houses, has always been an evolutionary process. As small-craft expert John Gardner noted, "There isn't anything really as a new boat. New boats are all combinations of old boats worked over. An occasional feature may be new, but that is all." When the members of the Oquossoc Angling Association searched for a fishing boat, they looked to boats they had known in other regions that served needs similar to their own. In an account of a trip to the Rangeley Lakes in 1869, R. G. Allerton, a founding member of the O.A.A., wrote:

> As to boats, those of the Saranac and Long Lake are models of beauty and speed. Maine is much behind the Adirondacks in this matter, but she will soon improve, as a contract is now being filled for a number of new boats, to be constructed by an experienced builder, and there will be some improvements made even over the Saranacers.

It is no surprise that Allerton looked to the Adirondacks in a comparative and competitive way. Many of the early Oquossoc members were New Yorkers who had previously summered and fished there, and it was natural for Allerton to compare not only boats, but the fishing and scenery in the two regions. Yet when it came to commissioning a boat, these men knew better than to mimic the Adirondack guide-boat. Just in the midst of its transformation from square stern to double-ender, the sleek—almost delicate—guide-boat had been honed to a minimal weight for portaging between the little lakes of upper New York State. Conditions in the Rangeley Lakes were otherwise, and a different boat was needed.

The most definitive statement of what the angling club chose comes as reminiscence from Frank C. Barker, a guide, steam-launch owner, and hotel proprietor who was a mainstay of the region for years.

> I think one of the best investments I ever made was parting with sixty hard-earned dollars for one of the first six Indian Rock boats that were built at Rangeley by Mr. Ball, the old-time superintendent of the Oquossoc Angling Association, and Luther Tibbetts, taking for a pattern a boat brought from Ogdensburg, New York. They were lap-streaked cedar boats, shaped much like a canoe, being pointed at both ends, and about sixteen and one-half feet long, and which are now so common all over the lake region, the model of which has changed but little.

Although Barker mistakenly referred to Mr. Ball as the superintendent of the Oquossoc Angling Association, there is an aura of authenticity about this statement, as there is about the volume it is drawn from. In *Lake and Forest As I Have Known Them,* F. C. Barker's recollections of Rangeley were not based on having passed through once or twice, but on having made a living there over the course of a lifetime.

The boat from Ogdensburg, New York, after which the Oquossoc Angling Association modeled theirs, was likely an early form of what became known as the St. Lawrence skiff. The St. Lawrence River is encrusted with islands as it runs into Lake Ontario, and here at the border between America and Canada, the Thousand Islands became New York's second great destination for fishing and summering. From the 1850s onward, guides rowed gentlemen out for bass and muskellunge, and their families for picnics and pleasure. The boat they used was gradually evolving in workshops along the river, in Ogdensburg and Clayton; it was lapstraked and cedar-planked, and could be 16 feet in length or more than 20. It was a long and shapely boat, but seaworthy and easy to row, no small matter on a river twelve miles wide in places and dangerous in stormy weather.

That the Oquossoc Angling Association did take as its pattern a boat from Ogdensburg and begin the construction in 1869 is borne out in several ways. As the first Rangeley visitors knew the Adirondacks, so they also knew the Thousand Islands' waters and fishing and saw it as paralleling the environment of the Maine lakes. At the rear wall of the O.A.A. superintendent's office hangs a framed photograph dated 1867 of the association's grounds and waterfront at Indian Rock. Alongshore rest three "Indian Rock" boats that would be identified as Rangeleys today, joined by a bateau-like craft akin to the working boat of log drives. Possibly these craft are early St. Lawrence skiffs, imported for fishing and to serve as the pattern for the association's own fleet of fishing boats. Further confirmation of F. C. Barker's recollection that the earliest Rangeleys were built in 1869 comes from the first account book of the Oquossoc Angling Association. In November of that year, the association made payments totaling three hundred dollars to Mr. Ball for "boats," and also compensated Mr. Baker Tufts for a boat. Baker Tufts would continue as a boatbuilder in Rangeley for some time to come.

The advantages of the Indian Rock boats—soon known as Rangeleys—over what had preceded them on the lakes were considerable, in Barker's testimony.

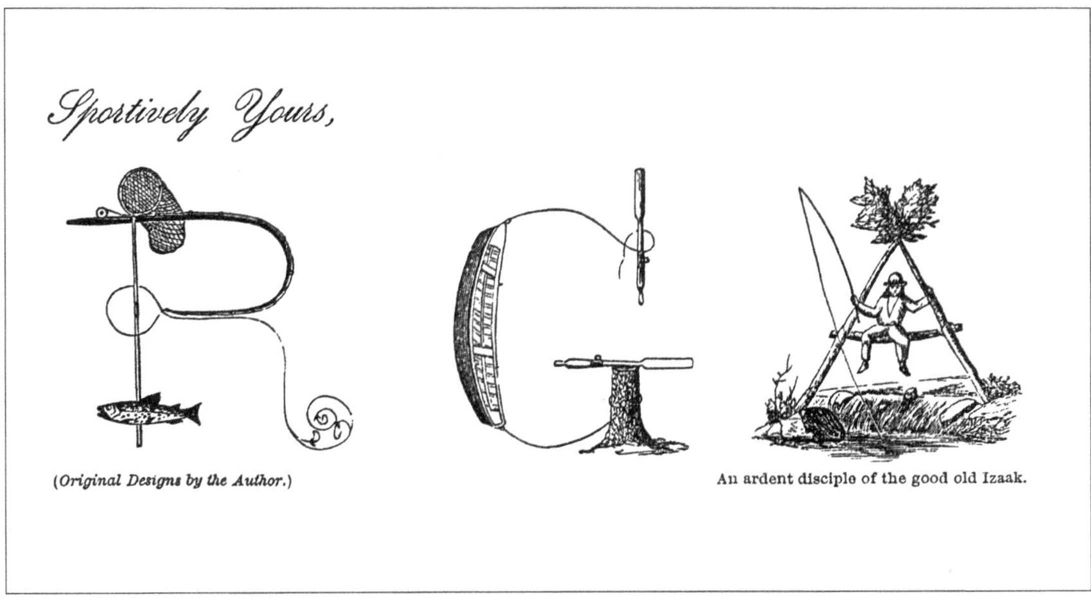

The initials of the piscatorially inclined R. G. Allerton, from his book *An Account of a Trip of the Oquossoc Angling Association to Northern Maine in June 1869* (Maine Historic Preservation Commission)

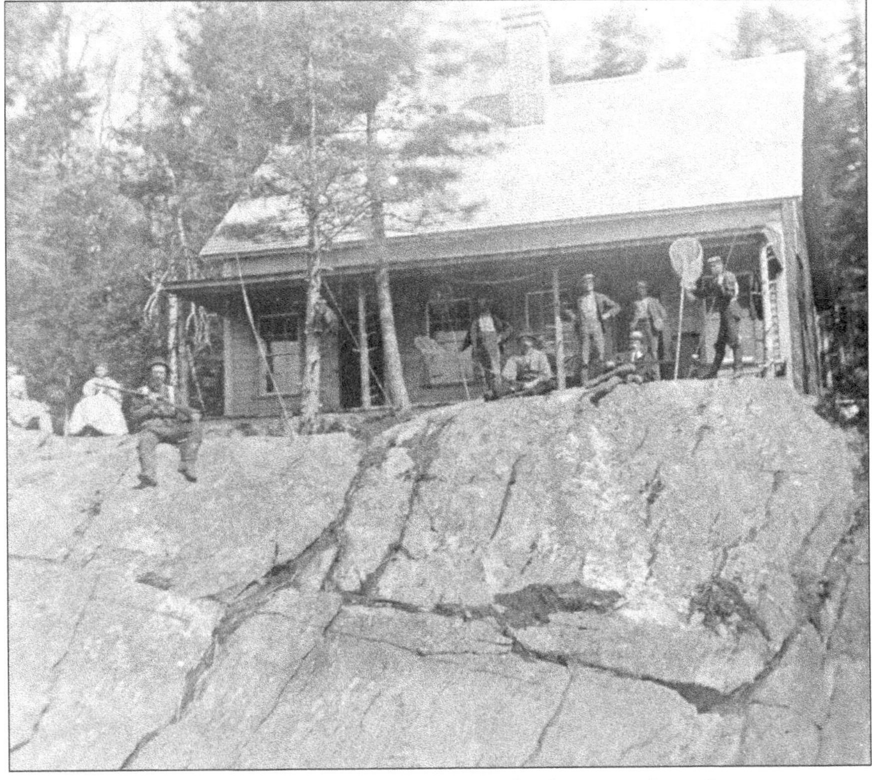

R. G. Allerton's lodge at Bugle Cove, Mooselookmeguntic Lake
(Maine Historic Preservation Commission)

The boats which the guides used on the Richardson Lakes were all quite heavy, and more or less hard to row; and although nearly all had sails, it was seldom we had a chance to use them to very good advantage, and generally more time would be wasted in trying to sail than if depending on our oars. With this boat I could make the pull from the Upper Dam to the arm of the lake and back in almost any ordinary day, and take about as much load as any of the large boats. When the old guides first saw the boat, they told me I had fooled away my money, for the boat would not stand either baggage or rough weather. In this they soon found themselves mistaken, for I could go in almost any wind that blew, and with a fair load could pull up past Hardscrabble, while they were obliged to find shelter at Saint's Rest. It did me good service for a few years, and then I sold it for what it cost me, and an Upton guide used it for his guide boat for more than twenty years.

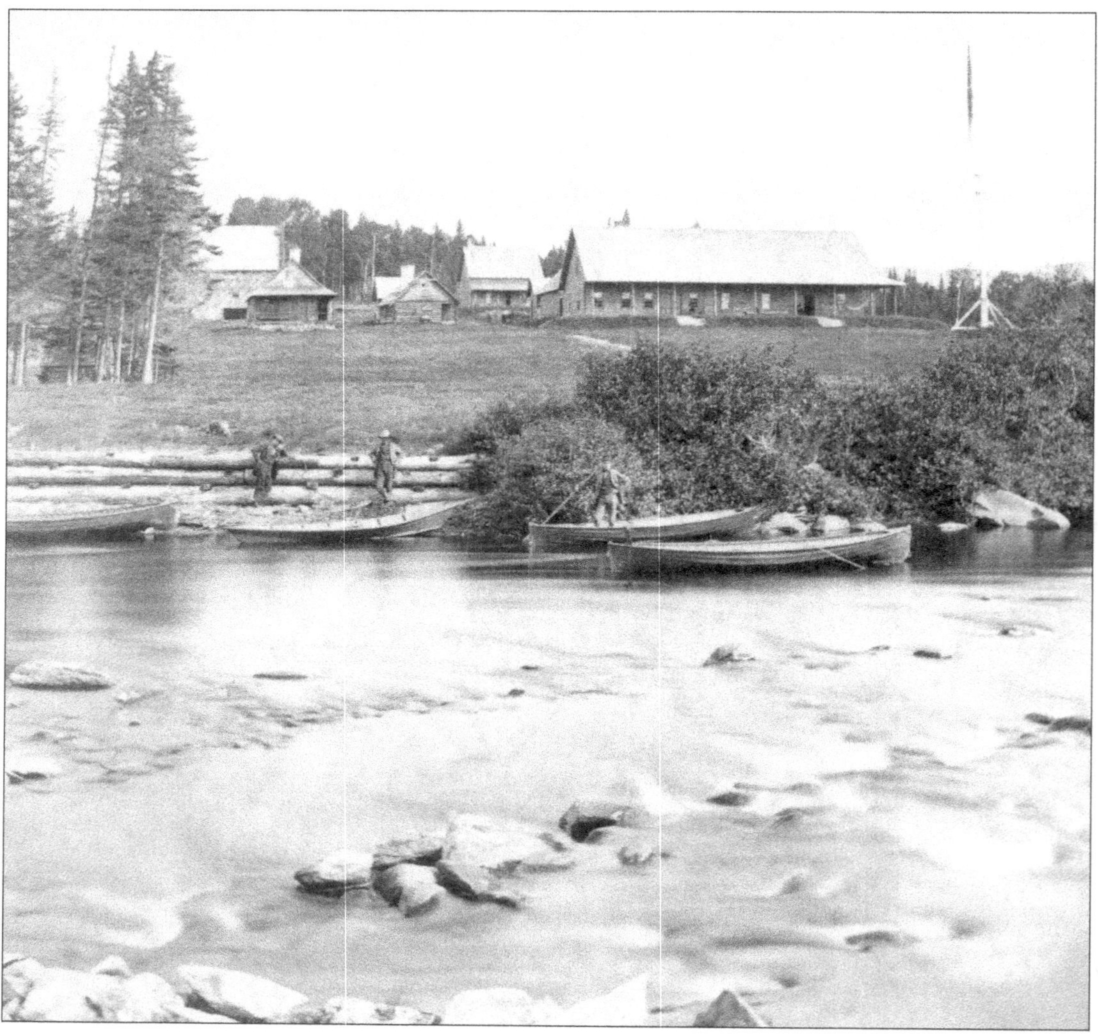

Early "Indian Rock" boats and an "Umbagog" boat at the Oquossoc Angling Association shore *(Maine Historic Preservation Commission)*

Ease in rowing and durability—two attributes not always combined in nineteenth-century pulling boats—were certainly what made these boats special on the Rangeley Lakes. Pulling ease and speed were a function of both shape and length, and the evolving Rangeley was long, narrow, and sleek. Its design was canoe-like with a sharp bow and stern, and its strength came from the lapped cedar strakes and many ribs within. When Harland Kidder began as superintendent at the Oquossoc Angling Association nearly sixty years ago, one of these early Indian Rock boats remained on the grounds. Mr. Kidder recalled it as comparable to later Rangeleys in size and construction, but fitted out with a seat snug up against the bow identical to the Rangeley stern seat.

What were the boats that natives, fishermen, trappers, and others used before the arrival of the Rangeley? A bateau-like boat is pictured in the early photograph of Indian Rock, and these shallow-draft lumbermen's vessels may have seen general use on the lakes in the mid-nineteenth century. Marshall Tidd, a Boston lithographer traveling the nearby Magalloway River and Lake Umbagog in 1861, used a boat he described as "in style about half way between a Swampscott Dory and a Batteau...about fifteen feet long, sharp at both ends." Tidd sketched these boats within his journal, and given his profession, the illustration may be quite accurate. What appears are lapstraked double-enders with the sheer of a Rangeley but with a crude, heavy, beamy quality about them. Perhaps this is the boat shown on the float at Indian Rock in 1867. Despite certain Rangeley characteristics, the Umbagog boats were, apparently, either overlooked or rejected when the Oquossoc Angling Association sought a prototype for its new fishing boat.

Other boats were on the Rangeley Lakes, as well. In late-nineteenth-century illustrations of the region, many open boats with high, square sterns are seen. There is a good chance that these are Whitehalls, the all-purpose pulling boat of the era that was used in every harbor and on every lake. It is certainly possible that some Whitehalls had been used on the Rangeley Lakes since the mid-nineteenth century. Finally, there is evidence that Adirondack guide-boats were occasionally seen on the rivers, smaller ponds, and lakes of the region. Recounting the events of an 1876 trip from the Richardson Lakes to Parmachenee Lake, writer C. A. J. Farrar notes picking up "a light 'Graves' boat, of the Adirondack pattern," for the journey.

The man who assisted in the construction of the first Rangeley boats at Indian Rock would soon be acknowledged as Rangeley's first boatbuilder. In 1872 the *Maine State Year-Book and Legislative Manual* noted Luther Tibbetts as both a wheelwright and boatbuilder, a designation which appeared until 1889, when Mr. Tibbetts's name disappeared from the listing. Luther Tibbetts's shop appears to have been an important place in the tiny community of Rangeley, and to have included, of all things, a bakery. This curious piece of information was reported by writer Edward Abbott, who wrote about the "Androscoggin Lakes" for

THAT TRIP TO MAINE.

June 2nd, -- 21st, 1874.

LAKES MOOSILLOMAGUNTIC, MOLLYCHOOKAMOOK, WEELOKENNEBACOOK, AND UMBAGOG.

Party.	Guides.
Mr. & Mrs. George F. Baker.	David Toothaker.
Mr. & Mrs. Fred. F. Thompson.	William Ellis.
Mrs. Mary Dock.	Amos Ellis.
Wallace W. Fahnestock.	Isaac Bradine.
Dr. Charles E. Huckley.	George Oakes.

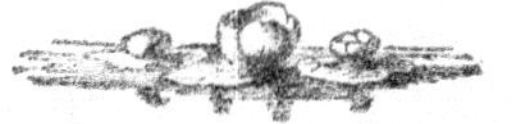

(Maine Historic Preservation Commission)

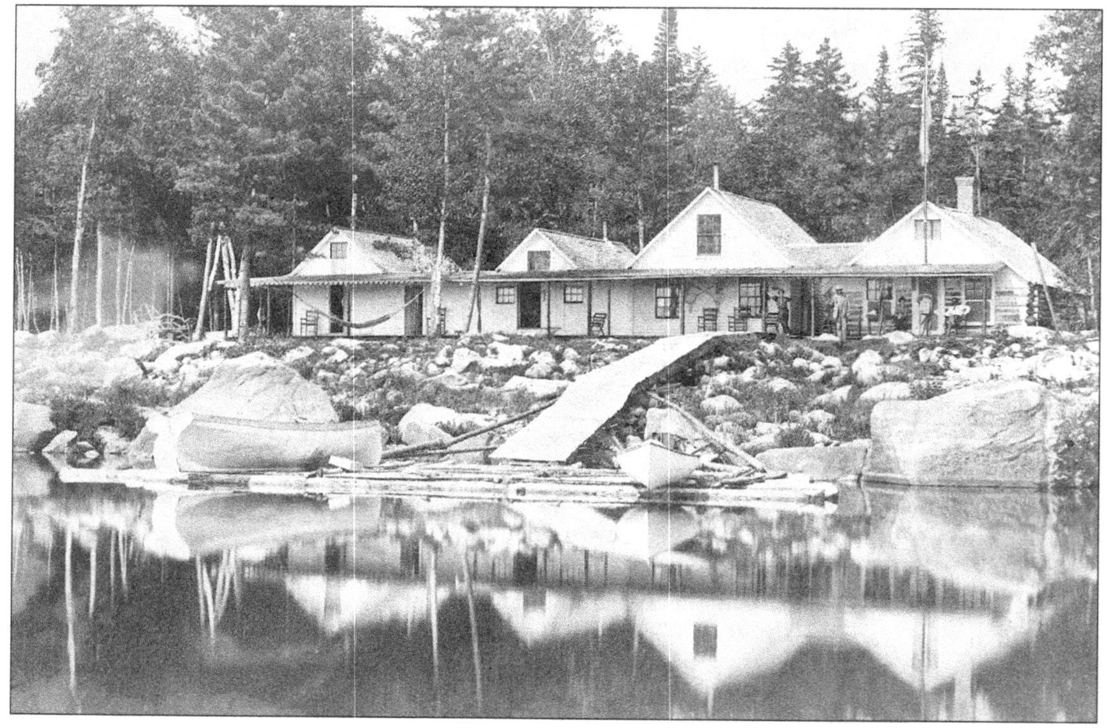

Boats from away on the float at Camp Bellevue, near Upper Dam, circa 1870–80
(Maine Historic Preservation Commission)

Harper's New Monthly Magazine in 1877. The story appeared one year after downtown Rangeley had been consumed by a fire which began in the Tibbetts shop.

> It is but fair, however, to explain that this backwoods bakery is the ingenious and accommodating device of Mr. Tibbetts, the boatbuilder, who, of a Saturday night, cleans out the furnace of his steam engine, and bids his neighbors bring thither their pots of brownbread and beans for a night of it.
>
> Mr. Tibbetts's "bakery" is not the only thing to his credit. His boats have carried his fame all over the lakes, and his shop is a place which every sportsman in the region makes a point of patronizing. He builds his craft something after the model of a "birch," first framing his streaks on a mold, and then strengthening the shell by a light knee-work within. The boat emerges from his hands at a cost of about fifty dollars, easily carrying six men, and easily to be carried by two or three. They are models of lightness, swiftness, and beauty.

While the writer embellished the carrying capacity of the Rangeley, his final description of the boat was unassailable.

In 1876 Baker Tufts was listed by officialdom as a boatbuilder, and the number of men making Rangeley rowboats grew to a fraternity of two. By 1890 H. W. "Hod" Loomis had also joined the trade. Ten years later, a correspondent for *New England Magazine* made his way to the sporting paradise of Rangeley and wrote:

> It is the seat of a large boatbuilding industry. Mr. L. H. Tibbetts, previous to the "Great Fire," established the fame of the "Rangeley Boat," built of cedar and ash, light, strong, and graceful. Baker Tufts and H.W. Loomis have slightly varied the model, and their rowboats are unrivaled.

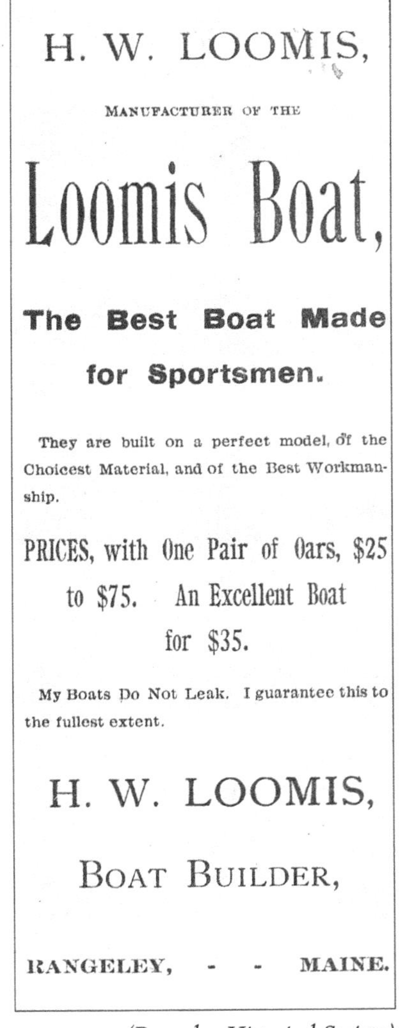

(Rangeley Historical Society)

Somehow, the journalist failed to mention a builder who had come to Rangeley some years before, and whose boats and legacy would outlast all others.

Rangeley Lakes.

VOL. I.   RANGELEY, MAINE, THURSDAY, SEPTEMBER 12, 1895.   NO. 16.

Rangeley Mineral Spring,

RANGELEY,   MAINE,

John B. Marble, Proprietor.

Acknowledged by all to be surpassed by no other water in curative properties. Prominent New York, Philadelphia and Washington physicians, are enthusiastic in their praise of the Rangeley Water. IT HAS RESTORED HEALTH AND HAPPINESS TO MANY. IT WILL DO THE SAME TO YOU.

SEND FOR DESCRIPTIVE CIRCULAR.

How Would YOU Like to Make a Catch Like This?

6 lbs., 3 1-2, 2 1-4, 1, 1. Taken by Miss Grace E. Hobart of East Bridgewater, Mass., in two hours' fishing.

A 5-pounder within five rods of the hotel. Taken by Mr. George Churchill of Brockton, Mass.

4 1-2 lbs., 3 1-2, 2, 1 3-4. Taken by Mr. C. R. Hoopes, a Philadelphia sportsman.

93 fish weighing 150 lbs. Taken in seven days' fishing by Mr. Walter B. Farmer of Arlington Heights, Mass.

ALL OF THESE HAVE BEEN MADE SINCE MAY 10, 1895, WITHIN SIGHT OF THE

Mooselookmeguntic House

HAINES LANDING, - - E. B. WHORFF, Proprietor.

SPECIAL RATES TO SUMMER BOARDERS.

Telephone Connection to All Points. - Daily Boston Mail.

Pleasant Island Camps,

LAKE CUPSUPTIC.

One of the most celebrated localities in the Rangeley region is now open to the public.

NOTED for LARGE and NUMEROUS CATCHES.

Everyone Catches Fish Here---There Are No Blanks.

Billy Soule's Line of Steamers Run to and from All Points on the Cupsuptic and Mooselookmeguntic Lakes.

BILLY SOULE, Proprietor.

P. O. Address, Haines Landing, Me.

The RANGELEY LAKE HOUSE,

AND COTTAGES,

RANGELEY, MAINE,

JOHN B. MARBLE, : : PROPRIETOR.

Accommodate 200 People.

Telegraph and Telephone

Connection to All Points.

Daily Mails to and from Boston.

SEND FOR ILLUSTRATED CIRCULAR.

MOUNTAIN VIEW HOUSE.

OUTLET OF RANGELEY LAKE.

Adjacent to the South Bog and other famous fishing grounds. This popular hotel is too well-known to need further description. The hotel is now supplied with the

Purest of Spring Water.

To those unacquainted with the house or surroundings, the proprietors will be pleased to give all desired information, and promptly answer all inquiries. Address,

KIMBALL & BOWLEY, RANGELEY, MAINE.

Phillips Hotel.

Sportsmen, Tourists and Others, returning from Rangeley via the Phillips & Rangeley R. R., have

One Hour for Dinner

in Phillips.

The Phillips Hotel is but one minute's walk from the station. The table and service are the very best.

A. L. MATTHEWS, · Proprietor, · Phillips, Maine.

Hotel Printing!

The Rangeley Lakes is prepared to do the finest class of work for Hotels. Anything from a card to the most elaborate half-tone or color work. Letter Heads, Envelopes, Menu Cards, Booklets, or Circulars.

Rangeley Lakes Publishing Company,

RANGELEY, ME.

[We hasten to use this space, before the eye of some enterprising advertiser lights on it]

The Barrett Shop

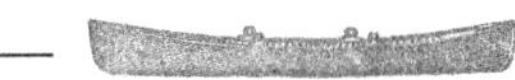

The town of Weld is no great distance from Rangeley. It is from Weld that two brothers, Charles and Thomas Barrett, came to Rangeley to build boats near the end of the nineteenth century. Soon after, the name Barrett was almost always said in the same breath as Rangeley boat. The best information on the Barretts and their boats is not culled from books or newspapers, but comes instead from the recollections of the late Edwin Barrett of Orrington, Maine. Edwin was Thomas Barrett's youngest son, and his earliest memories of the boat shop dated from 1920, when he was eight. In the 1930s Edwin spent five years working with his older brother Frank in the family business. Their father Thomas had started in at woodworking with incredible disadvantages. At fifteen, he had contracted meningitis, which left him sightless in one eye and deaf. Thomas Barrett also worked with little help from his left hand, which was crippled.

> My father built the first [Barrett] Rangeley boat. And Uncle Charles, being in better shape to talk with the public and so forth, he took the plans of the boat and went to Rangeley and started building 'em. Then he bought a house for my father to live in and, of course, father worked for him. And then father had a little shop of his own in which he repaired furniture.

The Barrett shop, which the brothers had established in the 1880s, still stands in town at the shore of Rangeley Lake. A simple two-story building, it retains the character of a workshop, although it is one no more. One can easily imagine the sign that hung over the door for years: RANGELEY ROWBOATS FOR SALE. NEW AND SECOND-HAND. REPAIRS. Edwin Barrett's memories of the shop date from its last decade under the elder Barretts, but the seasonal rhythm and work must have been little different from those of the shop's early days.

Well, Uncle Charles worked in the front part of the shop. My father worked in the back part of the shop. When I was a little fella I used to go down there, stay with my father. He'd set me on the bench, put something in the vise for me to work on and then—of course, I had to tap him on the shoulder if I wanted any information. And I used to play in a big pile of shavings at the end of his bench.

In the summertime we got out the lumber and we peeled it and stuck it outside. Then in the fall, we took the lumber and put it under cover.... We had pine for seats and the deck was pine and the oak for the rails and then we had spruce for the floor. It had a separate floor in there to keep you from walking on the ribs.

In the fall or in the winter when they start building boats, my Uncle Charles marked out all the planking and sawed it. And then it went in to father's bench and it was beveled. Part of the bevel was put on by machinery, by saw, it was a real sharp bevel. But the rest of it was all hand-planed for the bevel. Then after it was planked, it went to my father and he put the deck in. He put the rails on the side, and then it went upstairs and it was ribbed. It was ribbed with half-inch or five-eighths oak ribs, half round. Then after it was ribbed, it came back downstairs to my father and he put the cutwater on.

But Uncle Charles did most of getting the planking ready and he planked most of the boat. The rest of it was done by my father. And my brother was upstairs when he was old enough to rib 'em. And then when I went up there, I got all the ribbing.

They were all painted with white lead and oil inside and green paste on the outside. I think Uncle Charles and Frank did most of the painting.... The final coat of paint was put on, they were taken out of the shed. And then in the spring when they were ready for sale, they brought 'em in and put an extra coat of paint on them inside and outside, so they were fresh.

While the boatshop and its management kept Charles Barrett occupied year-round, the slack months of summer and early fall found Thomas Barrett in the shop behind his house on Cross Street. Here he mended furniture for Rangeley's matrons and turned out a remarkable array of wooden items. Barrett crafted double-runner sleds, roll-top desks, wheelbarrows, pony carts, sleighs for babies and ones for their dolls, a variety of children's games, and even bows and arrows.

The ability of the Barretts had been acknowledged immediately on their arrival in Rangeley, and found expression not only in the boats and other wooden items they turned out, but also in prose. In 1886 the Dickson family of Philadelphia purchased wooded Ram Island in Rangeley Lake and began construction of a grand summer house and gardens. Inevitably, lawyer Frederick Dickson came to rely on Charles Barrett and wrote of him in *And The Wilderness Blossomed,* a volume on his island home and grounds.

Some of these men are superb mechanics, the old-fashioned kind, who not only know all branches of their own trade thoroughly, but also much of kindred trades. I brought with me the design of a corner

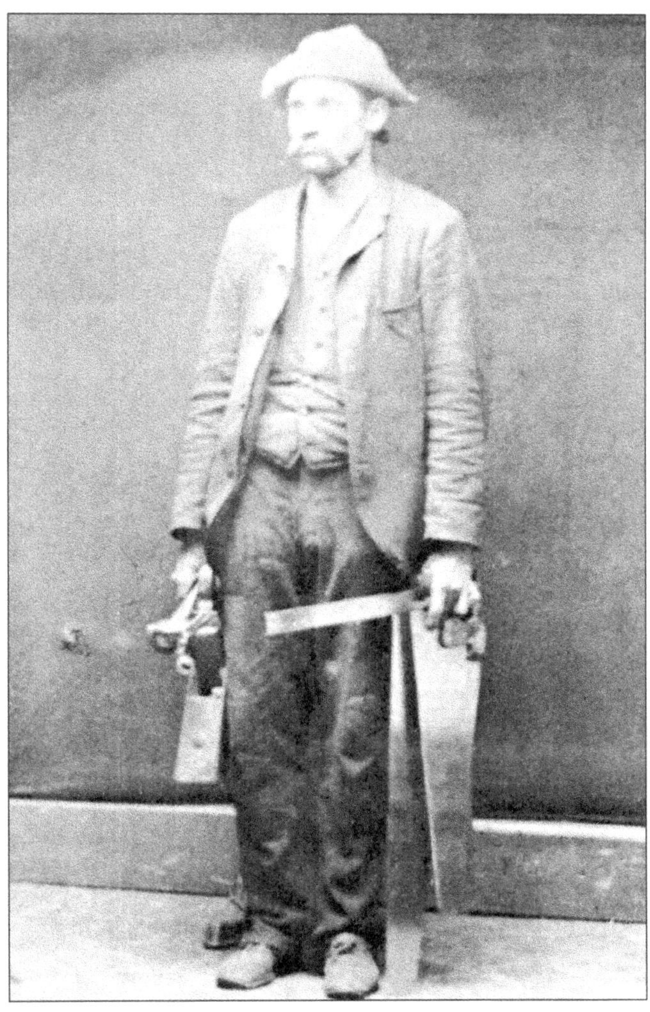

The quiet one: Rangeley boatbuilder Thomas Barrett
(Rangeley Historical Society)

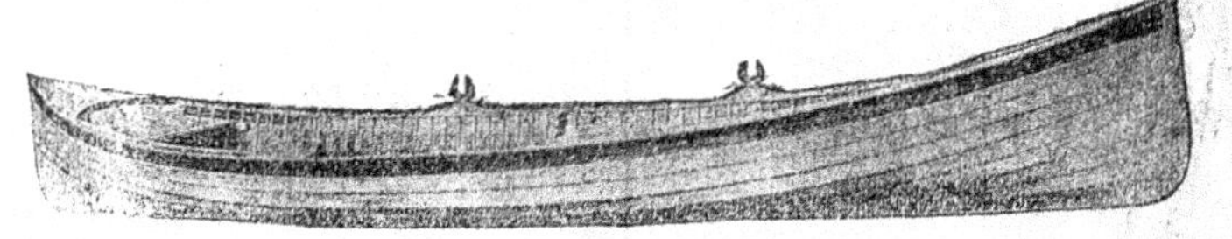

(Rangeley Historical Society)

closet to be built in the dining room, copied after an antique model containing those little diamond panes of glass which are so attractive. I showed the plans to Charlie Barrett, and asked him if he could build it. "I think so," he said musingly, "I haven't any plane to work out those styles, but I guess I could do it." He did not appear at the island the next day, and I found out later that he had spent the time in a blacksmith shop making and tempering a plane to correspond with my drawing. That these people should be so skillful in the management of metals with actually no training at all, is no less than marvelous to me.

It has been a common saying amongst us, that if Harry Furbish could not get what you wanted, Charlie Barrett could make it. It is true, too—absolutely so.

(Rangeley Historical Society)

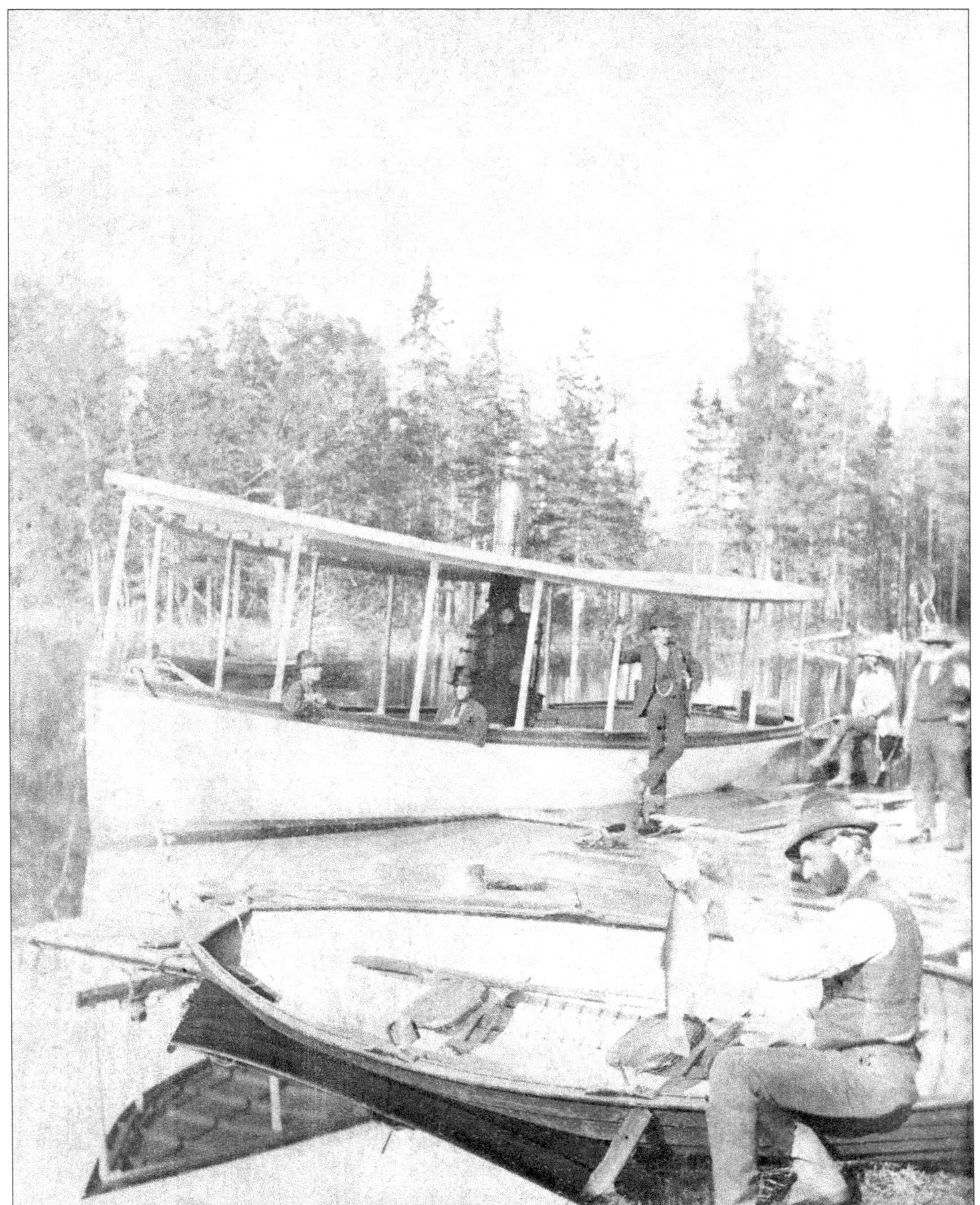

The Barrett-built steamer Oquossoc, circa 1873–78
(Maine Historic Preservation Commission)

Charlie Barrett would not only furnish the Dicksons with Rangeley boats and corner cupboards, but also a small steam launch. Steamers were common on the lakes for transporting visitors and pleasure cruising before the advent of the gasoline engine, and the Barrett shop produced at least two. All that is known about the steam launch *Oquossoc* comes from *The Islanders,* Elizabeth Foster's book about her grandparents' times on Ram Island. The Barretts apparently built the vessel with a copper-lined boiler, pressure gauges, and safety valves, and outfitted her with a striped canvas awning and cushioned seats.

Frederick Dickson's notion of Charlie Barrett as the resourceful old-time mechanic able to turn his hand to most any trade was apt. At least one and, very possibly, other innovations that began to appear on the Rangeley boat in the late nineteenth century can be attributed to the Barrett shop. These innovations came to distinguish the Rangeley as a distinctive, regional boat, something more than just a simply built St. Lawrence skiff.

The most notable of these innovations was the Barrett oarlock. Designed by Charles Barrett, it was an extremely functional round socket type. The oars were held by a metal collar fixed between the horns, which allowed a guide to feather them while rowing. Best of all, the leathers on the oars had shoulders, so that the oars could be dropped to change a fly or play a fish without their loss overboard. The St. Lawrence skiff's pinhole oar locks kept the oars secured to the boat, but didn't allow for feathering and must have been the bane of many a fishing guide's existence. The Barrett oarlock was cast outside of Rangeley but machined in the shop, a task that Edwin Barrett performed during his years there.

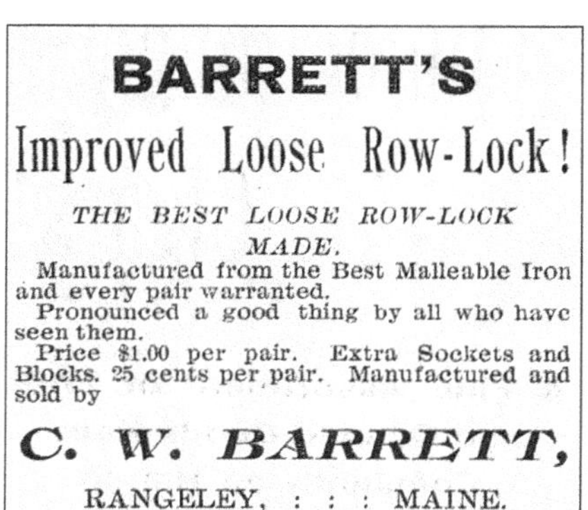

(Rangeley Historical Society)

The Rangeley's round seat was another characteristic of this boat alone. A single piece of pine, sawn round and slightly hollowed, was placed atop and at the middle of the thwarts. Simple and ingenious, the seat assured that all in the boat sat dead center on the thwarts, giving the advantage of perfect balance to the rower. The thwarts, too, gave evidence of intelligent design. They were built up at the sides, fashioned to keep the paraphernalia of fishing beside the fisherman, and not on the floor of the boat. Spare leaders, flies, perhaps a pipe and matches, could all be strewn about, yet were contained in the recess of the thwart. The round seat and recessed thwart cannot be attributed to a particular Rangeley boatbuilder, but Thomas Barrett's son had a hunch that the seat design came out of his family's shop. He speculated that a guide came to the Barretts in search of a way to keep sports and sightseers in the center of the boat, and his father and uncle responded with the round pine seat. Given the Barretts' reputation for skilled engineering in wood, the claim may well be true. As Edwin Barrett said, "They could do anything if it concerned wood and some iron—they could make anything that anybody had any desire for."

The Barretts' arrival in Rangeley could not have been better timed. During the 1880s the sportsmen and tourist trade there burgeoned, grand hotels were built, new sporting camps were established, modes of transportation were increased and updated, and publications wrote frequently of the Rangeley Lakes as a holiday destination. *Harper's New Monthly Magazine* had featured a heavily illustrated article on "The Androscoggin Lakes" in 1877, and the Boston & Maine and Maine Central railroads published a travel guide to the region in the same year. Charles Farrar of Boston put out *The Androscoggin Lakes Illustrated* annually in the next decade, and *New England Magazine* would tour and write of the Rangeley Lakes in 1900.

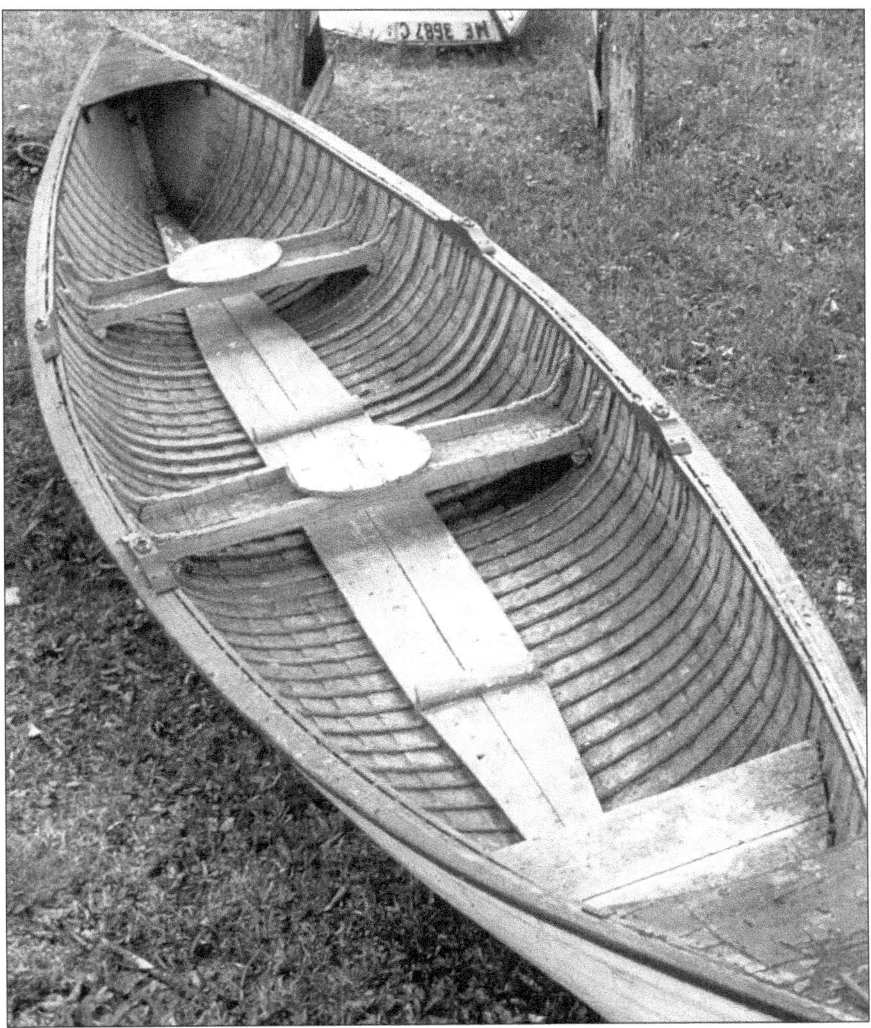

The Rangeley's round seat and thwart built for fishing storage
(Michael Chaney, Maine Folklife Survey)

The scale of tourism at the lakes was considerable. On Rangeley Lake, both Mountain View House and the Rangeley Lake House could accommodate nearly one hundred summer guests. New private camps owned by professional men from New England and the Mid-Atlantic states appeared at the shore in more remote locations, and still deeper in, on Mooselookmeguntic's south end and around Loon and Kennebago lakes, sporting camps dedicated solely to fishing and hunting were found.

For most of the era in which camps and hotels thrived, railroads were the means of transport to the fishing regions and the most active promoters of these destinations. Indeed, camps and railroads maintained a symbiotic, if paradoxical, relationship. The juxtaposition between growing industrialism and pastoral America and their disparate value systems

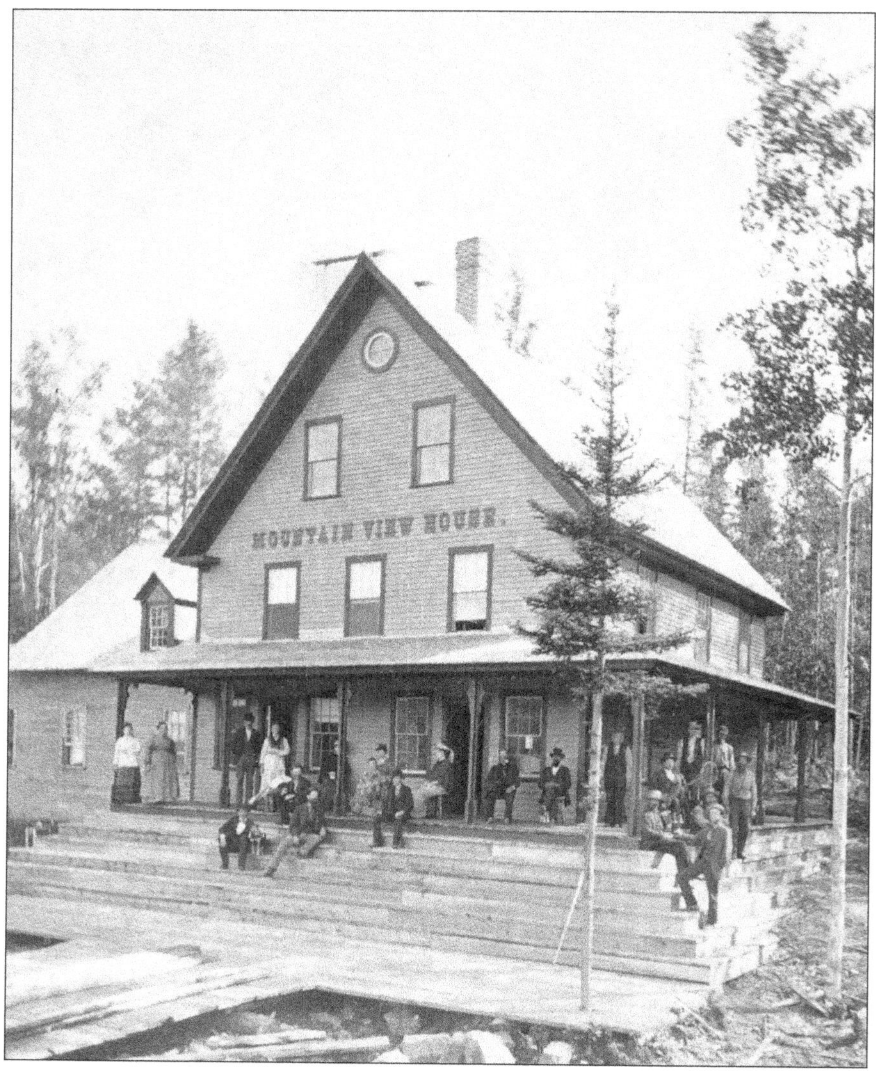

Mountain View House, Rangeley Lake Outlet, circa 1870–80
(Maine Historic Preservation Commission)

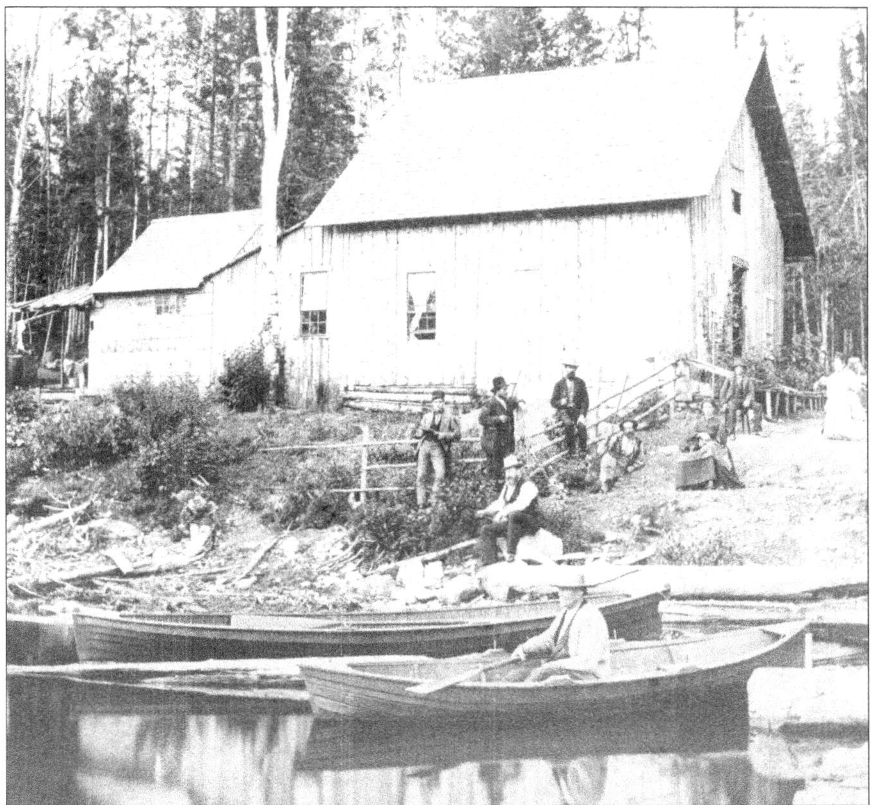

Soule's "Camp Henry" and Rangeleys, circa 1873–78
(Maine Historic Preservation Commission)

was a central theme in mid-nineteenth-century thought and writing. The theme played itself out candidly in the wilds of Maine where urban professionals and businessmen arrived via steam engine, the emblematic agent of industrial change, to partake of traditional pursuits in the woods. The very technology that made this pleasure possible irrevocably changed the wilderness environment these gentlemen sought. What had been inaccessible became accessible, what had been wild became less wild.

The railroads brought sportsmen and their families to the Rangeley Lakes, or nearly so. From Boston, one could reach Farmington, the seat of Franklin County, in a day. From there, the narrow-gauge Sandy River Railroad carried passengers along the river valley as far as Phillips, where a stagecoach completed the final twenty or so miles to Rangeley. It was a two-day trip. By 1891 Rangeley had direct train service from the east. Traveling to the region from points west, one employed a similar mix of rail and stage, in addition to a steamboat passage. In this wilderness, steamers were the logical means of transport between the lakes, and numerous ones—many with Indian or mock-Indian names—plied the waters. In the early 1980s Rangeley boatbuilder Herbert Ellis described the routine and its logic from memory:

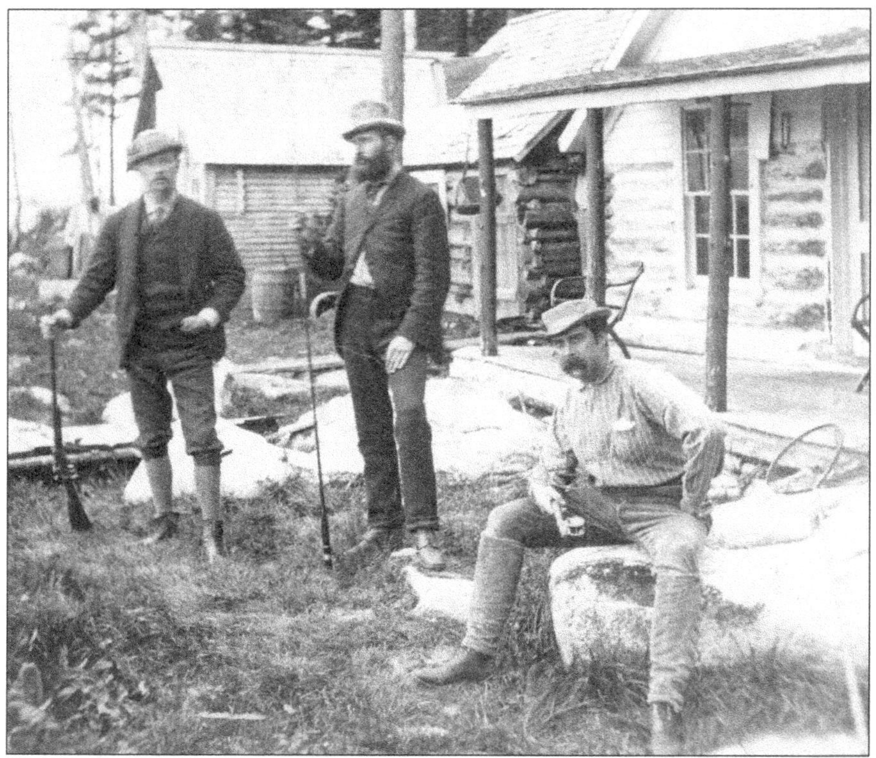

Victorian "sports" at Camp Bellevue *(Maine Historic Preservation Commission)*

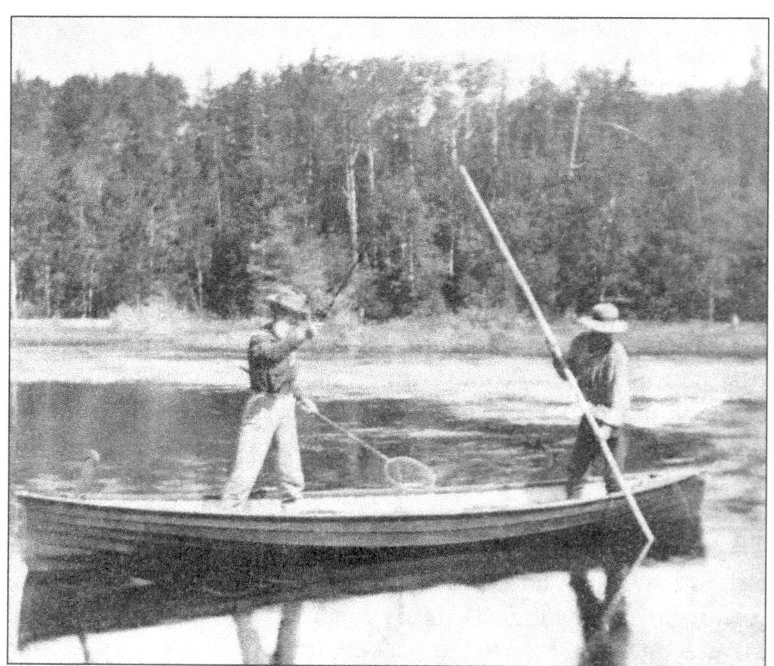

Fly-fishing with a guide's assistance, circa 1873–78
(Maine Historic Preservation Commission)

Used to have steamers run all these lakes, and people come in by train. They would come in South Rangeley station and it had a spur track that ran right out on a big wharf, and a steamer'd pull in there and they'd back up and unload the luggage, trunks and everything, put them on the steamer, and the people would get on and they would go to Pickford's Camps and Rangeley Lake Hotel and then to the town dock. Dropping off passengers, luggage, and mail at all the stops. Then going down to the other lake, they would stay on the train until they got to Oquossoc, and then a horse-drawn stage would take them down through to Mooselookmeguntic Lake. So really, the easiest transportation was by water.

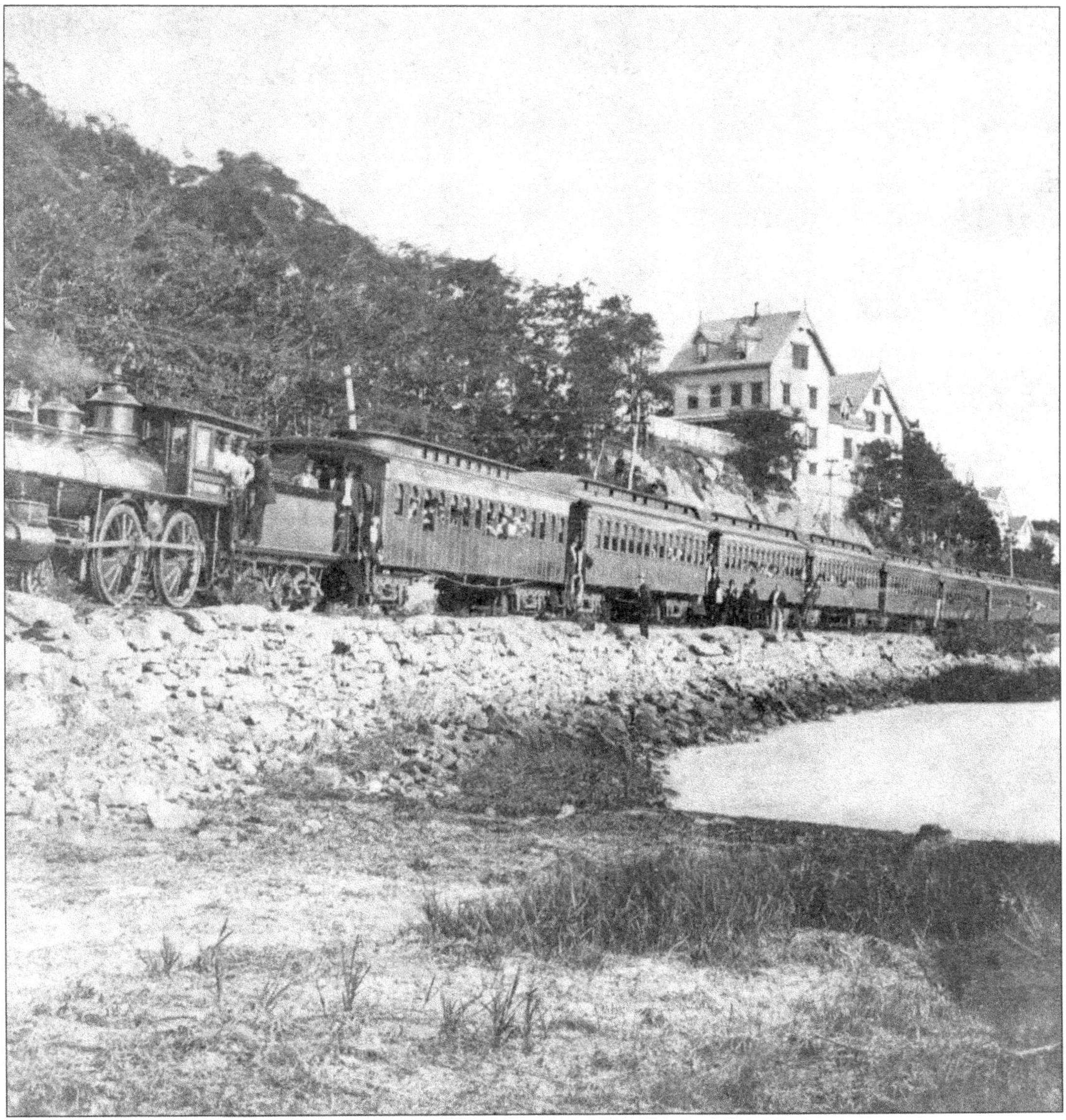

The Maine Central Railroad between Farmington and Rangeley, circa 1873–78
(Maine Historic Preservation Commission)

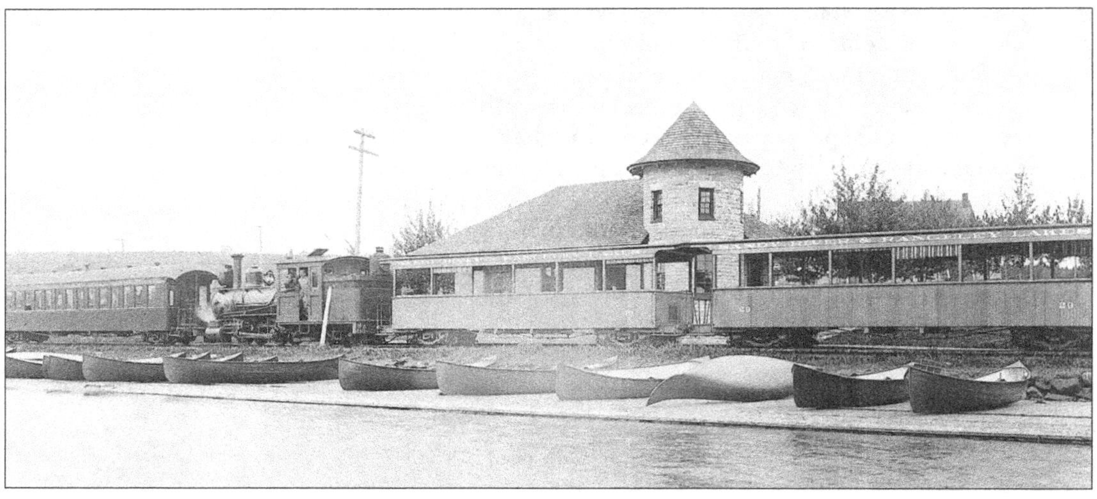

Rangeley Station and Rangeley boats *(Rangeley Historical Society)*

Parlor Car Service

via South Rangeley, to Rangeley, and Loon Lake. Boats make close connections at South Rangeley with all trains to and from Portland, Boston and New York. Parlor car to and from South Rangeley.

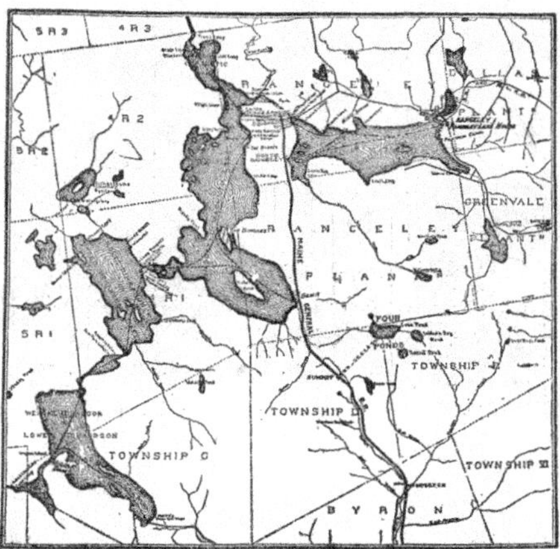

(The Rangeley Lakes)

Sleeping Car Service via Farmington

To Mountain View, Haines Landing, Indian Rock and Bald Mountain

Sleeping Car between Farmington and Boston

Boat leaving Rangeley at 8.00 a. m. week days, 9.00 a. m. Sundays, connects at that point with train leaving Boston at 10.00 p. m. previous night.

Boat leaving Rangeley Outlet at 5.00 p. m. connects at Rangeley with train arriving in Boston at 5.10 a. m. next morning. Daily, Sundays included.

RANGELEY LAKES
STEAMBOAT COMPANY

TIME-TABLE

IN EFFECT JUNE 23rd, 1913

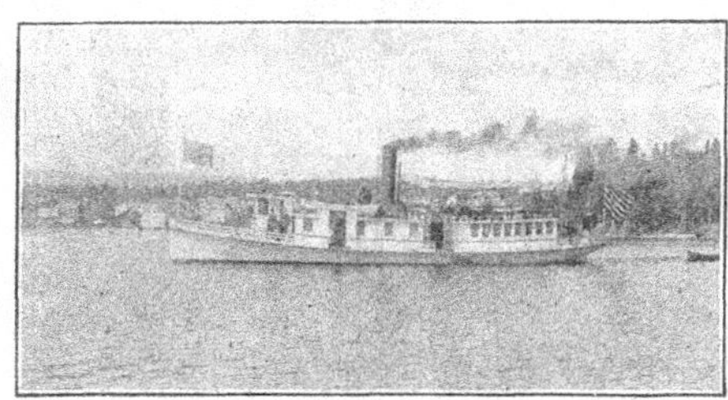

(STEAMER RANGELEY)

Through Parlor Car Service on all trains to and from South Rangeley.

Pullman tickets on sale at steamboat office on Rangeley wharf.

H. H. FIELD,	D. F. FIELD,
Pres't and Gen. Man.	*Treas.,*
Phillips, Me.	**Phillips, Me.**

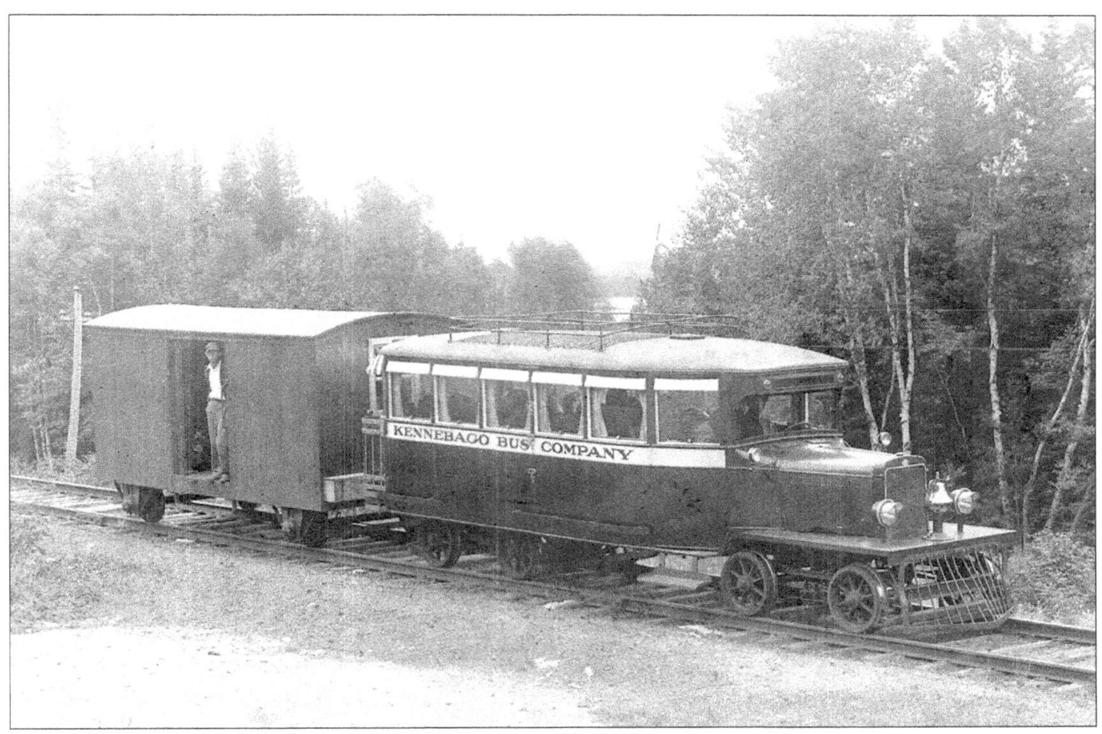

The Kennebago Bus Company traveled to remote Kennebago Lake *(Rangeley Historical Society)*

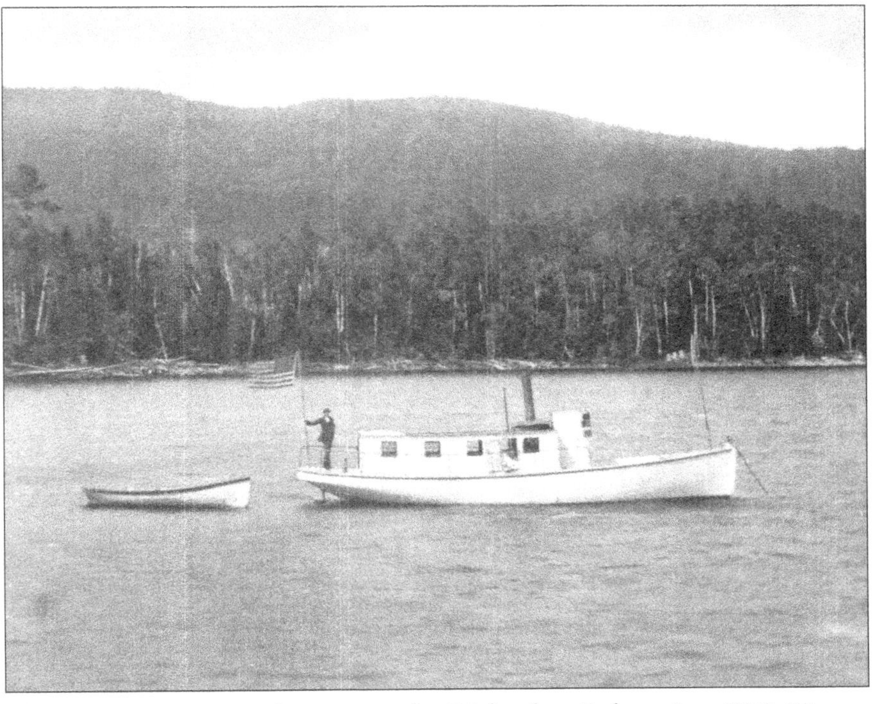

The steamer Henry B. Simmons on the Richardson Lakes, circa 1870–80
(Maine Historic Preservation Commission)

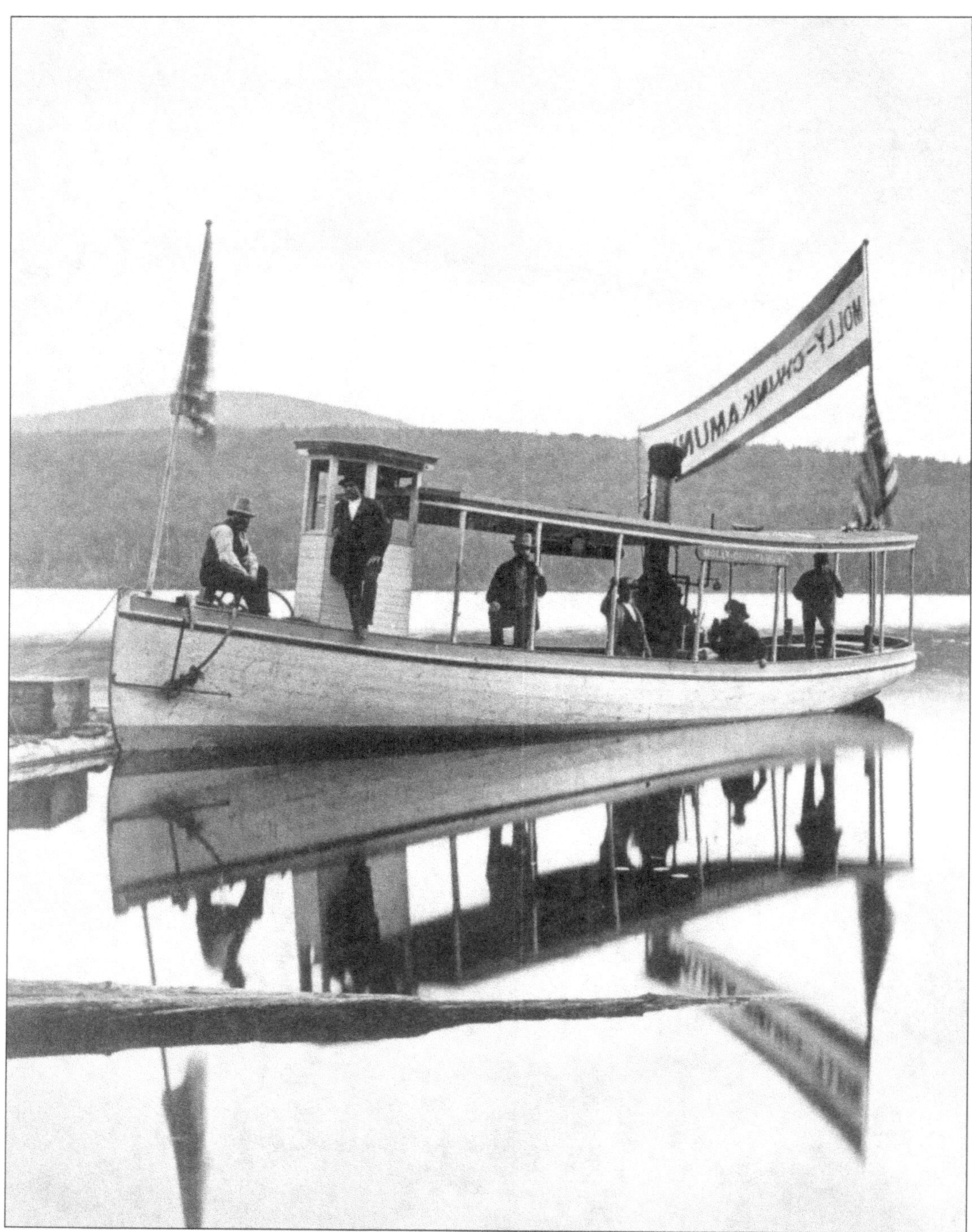

The diminutive steamer *Molly-Chunkamunk*, Upper Richardson Lake
(Maine Historic Preservation Commission)

From 1888 until 1929 Thomas and Charles Barrett provided their Rangeley boats to an expanding population of sportsmen, guides, and sporting camp and hotel owners. It is possible to closely follow the operations of the Barrett shop in the pages of the short-lived *Rangeley Lakes*. From 1895 to 1897 this small-town newspaper chronicled the doings of natives, sportsmen, and summer folk, with a distinct emphasis on the social life of Rangeley's seasonal visitors. Columns such as "Camp and Cottage" appeared regularly, and the paper published a "Weekly Register" of guests lodging at the Rangeley Lake House. Fishing news was not forgotten, either: "Walter Buckley, of Bristol, Connecticut, is much pleased with his 4¾-pound catch. Philbrick Guile, guide." The Barretts, as local businessmen linked to the town's summer trade, appeared regularly in the pages of *Rangeley Lakes*.

July 25, 1895, found the Barretts just concluding work on a special-order Rangeley.

> C. M. Barrett has just finished what is without doubt the handsomest boat ever built in Rangeley. It is for Harry Dutton who will use it on the Cupsuptic. The boat is seventeen feet long and is finished in natural woods. The planking is of cedar, save the top strip which is of oak. The gunwales are of maple. The decks are made of alternating strips of bird's-eye maple and a darker-grained wood. The flooring is of unique design and not an inch of the boat is finished in any other than the best possible shape. Mr. Barrett solemnly asserts that he is not a finished cabinet maker as well as boatbuilder, but an inspection of his work on this boat would seem to belie the assertion.

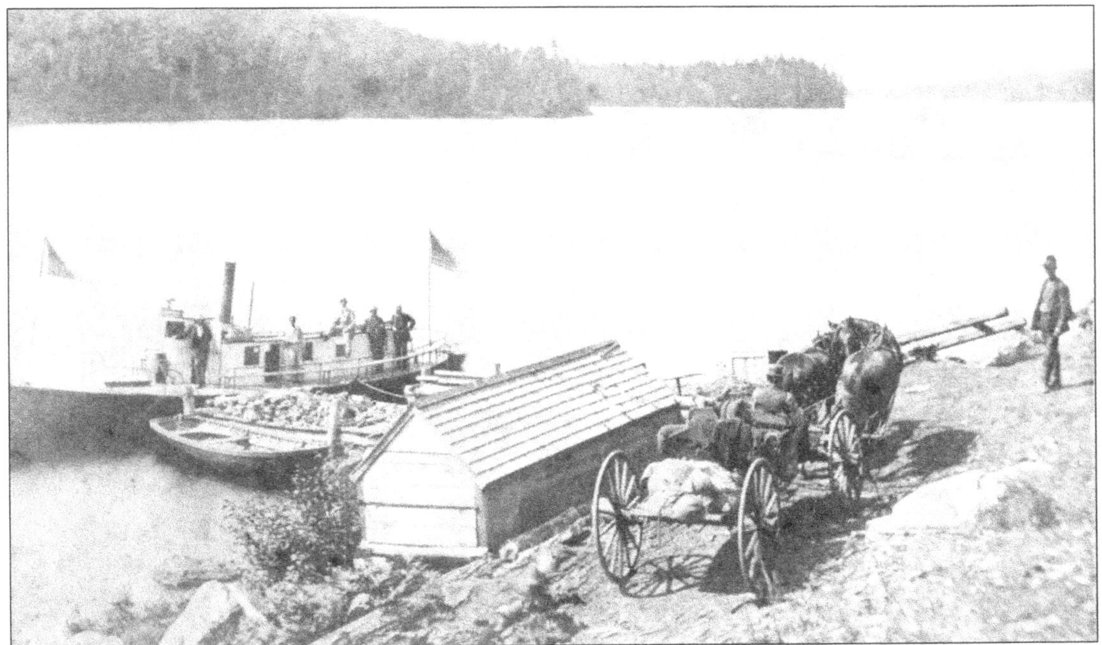

The steamer and buckboard transit, circa 1870–80 *(Maine Historic Preservation Commission)*

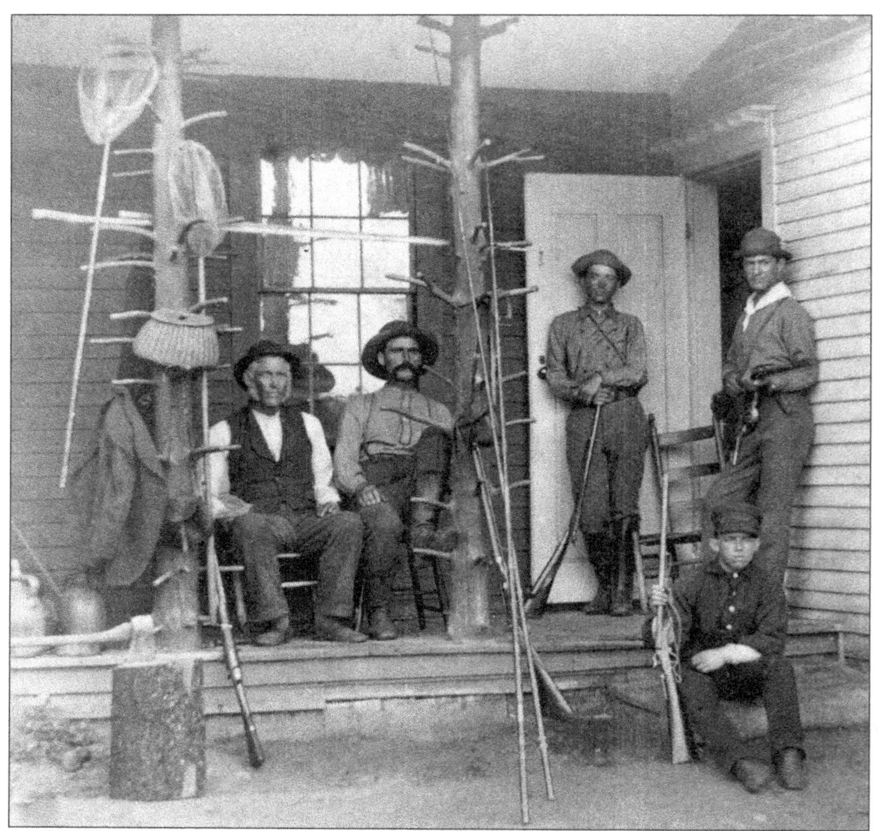

"Two old guides" 1874 *(Maine Historic Preservation Commission)*

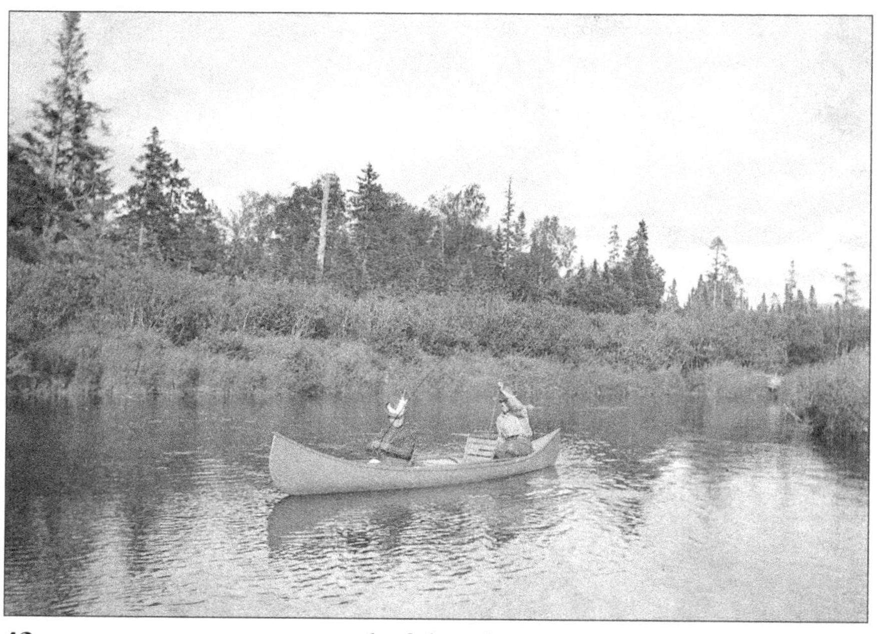

 Fly-fishing by canoe *(Rangeley Historical Society)*

Rangeleys now found on the lakes tend to have been built well into the 1900s. They are entirely functional, simply built, and always painted. It is useful to remember that there were patrons of these boats who lavished considerable money on construction, and that there were builders who, in attention to aesthetics and detail, crafted Rangeleys every bit as elegant as the more popular St. Lawrence skiffs.

In August 1895 *Rangeley Lakes* revealed that "C. M. Barrett is soon to commence the manufacture of canvas canoes," and by mid-September, the Barretts were well along in building their second.

> C. M. Barrett is making the second canvas canoe ever built in town. It is 16 ½ feet long and 33 inches wide. The ribs are of cedar and the planking of spruce. The first one he built is leased to parties at Billy Soule's.

The canoe became an important adjunct to the Rangeley boat throughout the region. While the boat served perfectly for lake fishing, a more narrow, nimble craft was needed for maneuvering in rivers like the Kennebago, where the fishing was often superb. Although it is known that

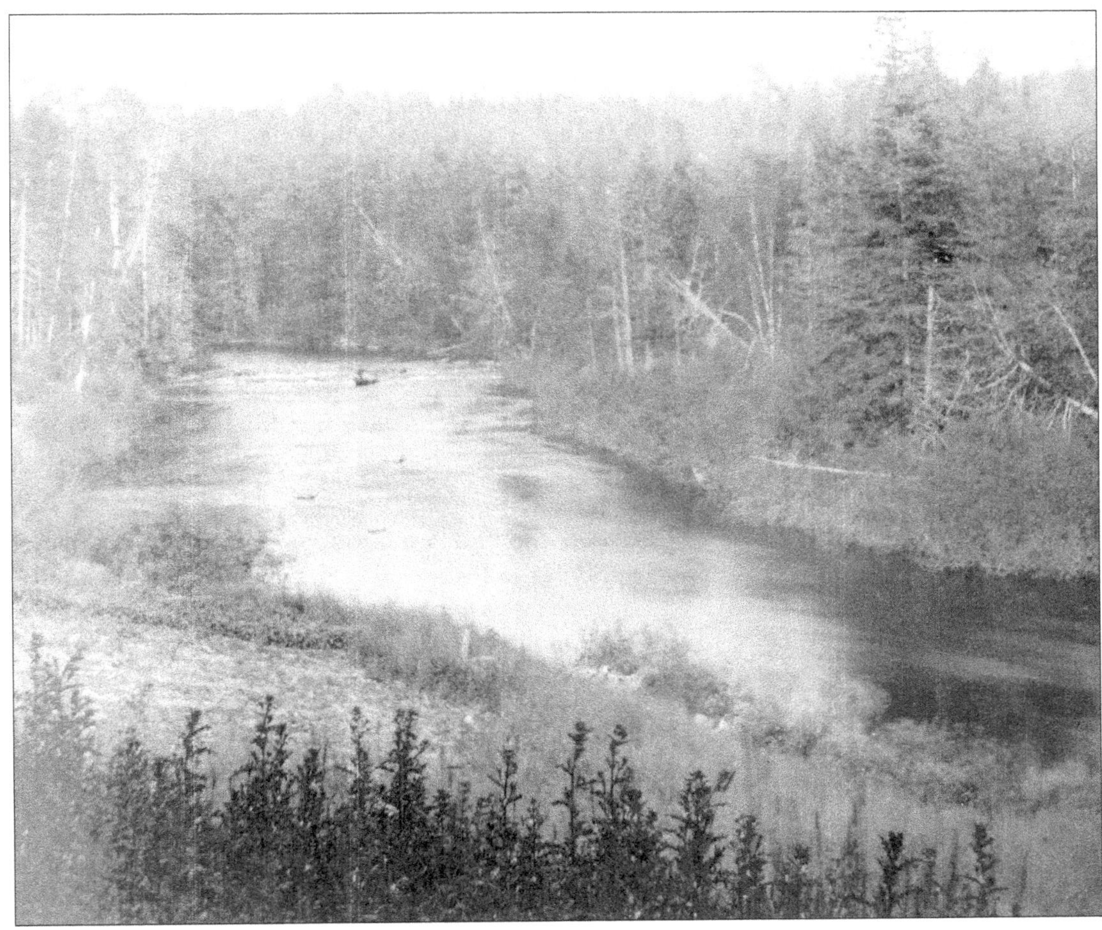

Kennebago Stream *(Maine Historic Preservation Commission)*

(Rangeley Historical Society)

the Barrett shop turned out both cedar and canvas-covered canoes, it seems that none have survived the years. In that same September entry, the newspaper added a second item worth remembering:

> Thomas Barrett has just completed a miniature rowboat 4 feet in length and 13 inches wide. It is in natural wood and very nicely done. It is built for Mrs. Lasell, of Whitinsville, Mass.

As the year wound down, Charles Barrett was applying to Rangeley's selectmen for permission to operate a steam engine in his new Main Street shop. The engine would power—via belts—the planes and saws of his trade as well as the shop's steam whistle, a welcome sound in town. He was also planning for the months to come.

> C. M. Barrett is busy preparing for his next season's boatbuilding. He built upwards of 30 last season and intends to get way above that number this winter.

Just prior to Christmas, in the December 12 issue of *Rangeley Lakes,* the editors profiled their advertisers—the businesses of Rangeley. E. T.

Hoar, who built split-bamboo fly rods was mentioned, as was Mrs. H. H. Dill, who sold millinery and fancy goods and tied dry flies. The newspaper also focused on the three builders of Rangeley boats to be found in town.

> Baker Tufts.
> He has, as you might say, grown up with the town, quiet and unassuming. Always the same, a friend of everyone. Mr. Tufts was the first boat builder, of any note, in this section, and has probably made more boats than any other person in town. And he still makes them, and good ones too.

Baker Tufts had been bending cedar into Rangeley boats for a long while. In 1869 he provided one for the fledgling Oquossoc Angling Association, and the *Maine State Year-Book* had officially listed him as a boat-builder since 1876.

Announcement.

IN making my annual announcement I wish first to express my thanks to my old patrons, both for their continuous orders and for the many kind words which they have spoken for my work. It is only through their appreciation that my boats have gained their present large sales and enviable reputation.

To old and new customers alike, I wish to say that, as in the past, my only aim shall be to produce such a boat or canoe as you may order. Of course I have my own ideas as to the best models to be used in different regions and for different purposes, and will freely give you the benefit of my experience in boat architecture. I am always glad to answer inquiries or make suggestions. A personal visit to my shop is invited at all times. But when you have made up your mind what you want, whether it embodies your ideas or my ideas, makes no difference. Just give me the directions and I will see that they are executed to the finest detail. Of course, work of this sort costs a little more than the ordinary models, but it will be cheaper than you can get the same elsewhere.

You will bear in mind that the very best material of every description grows right around Rangeley. Most builders have to charge you for freight. This saving, and the fact that I am equipped for the best and most economical work in every department, enables me to give you the best material, superior workmanship and any model you wish at a price which others cannot rival. It is not my aim to turn out a cheap boat or canoe but rather of the very highest quality.

I am prepared to build naptha launches, steam yachts, or any other water craft, of whatever pattern or design, as well as row boats and canoes.

H. W. "Hod" Loomis' annual pitch (*Rangeley Historical Society*)

H. W. Loomis.
The builder of the popular boat that bears his name. He gives one quality to his boats that is certainly in their favor. As he says, they do not leak. Mr. Loomis sells all the boats he can make.

Hod Loomis had worked at the trade for five years when this profile appeared, and would continue to do so at least until 1908.

The *Rangeley Lakes's* greatest enthusiasm seems to have been reserved for the last of the town's three boatshops.

C. M. Barrett.
Boat and canoe maker. Barrett's boats are found very generally distributed all through this region. He is a mechanical genius and can turn his hand to any work required, and is very pleasant to meet at his work or

Boats and Canoes.

In my own models I aim to build a boat of the least weight with the greatest carrying capacity, beauty, seaworthiness and speed. My prices range from $25.00 to $75.00. I build a thirteen foot row boat, painted, with pin row locks and one pair of oars, birch gunwales and oak ribs for $25.00, delivered on the train or steamer at Rangeley. A fourteen foot boat of the same pattern costs $30.00; sixteen foot, $35.00; seventeen and one-half foot, $40.00. The thirteen and fourteen foot boats have eight streaks and the larger ones nine. Loose row locks cost from $2.00 to $5.00 per boat additional. Extra oars are $2.00 per pair.

These same patterns in oil finish cost from $10.00 to $30.00 more, according to the style of finish. I do any kind of extra work that the customer may desire at a reasonable increase in expense.

My prices for cedar canoes, painted, are as follows: twelve foot, $25.00; fourteen foot, $30.00; sixteen foot, $35.00. Oil finish costs from $10.00 to $20.00 according to the style.

I usually ship boats without crate or burlap and have never heard of one being damaged at all; but the expense on either would be about $2.00.

I respectfully invite all interested parties to call upon me or write me and shall be pleased to answer all inquires promptly.

H. W, LOOMIS, Rangeley, Maine.

socially outside. He is ably seconded by his brother Thomas, and really, we think the brother can do a particular job a little the best.

The newspaper got it right on both counts, given what is known about the Barrett brothers from other sources. Deaf and partially crippled, Thomas Barrett was the consummate craftsman working behind the scenes. The brother whose name rarely appeared in the news, he had a special way with wood, making Rangeleys, a Rangeley miniature, and a host of other wooden items in his quiet world. Charles was the public Barrett, and his nephew Edwin confirmed both his uncle's mechanical proclivities and kindly nature.

For drills, the old-fashioned umbrellas used to have a rod for braces. So drills were too expensive, because every time you tried to drill a hole, every once in a while you'd break a drill. So my uncle came up with the idea that he took one of those braces and he filed it off kind of triangle shape or some other and we used those for drills. And boy, they really cut.

And one thing that I always remember; my Uncle Charles chewed some kind of tobacco, I can't remember, but it had tinsel [foil] wrapped up. And every time that anybody came in with a child, he always took a piece of that tinsel and made a goblet for the kid. And the kids used to love to come in, they wanted these little goblets.

A masterful engineer and sociable, too, Charles Barrett was undoubtedly the best possible promoter of the shop's boats. He also had other responsibilities. Charles Barrett served for a time as superintendent of the Rangeley Lakes Guide Association Fish Hatchery. Making certain that the lakes were replenished yearly with trout and salmon was an important task for a boatbuilder whose business depended on good fishing.

With the new year 1896, the *Rangeley Lakes,* seeking to increase its business, sponsored a contest among the fishing guides. The paper offered a painted Rangeley boat worth thirty-five dollars to the guide who sold the most newspaper subscriptions above twenty-five to sports and other visitors in the region. The winning guide could choose his boat from either the Tufts, Loomis, or Barrett shop.

While the guides scrambled for subscribers, Charles Barrett was working on a promotion of his own, as the *Rangeley Lakes* duly reported:

Hustlin' For the Boat.

The offer which RANGELEY LAKES made to the guides of Rangeley and vicinity has by no means been forgotten.

Every day or so a guide drops in with one or more new names and asks to have them credited to him in the contest. Some guides are writing to their sportsmen friends who will not be able to come this way the present season. In this way our list has received many valuable additions, and is growing every day.

Our original offer was as follows:

To the Guide who brings us in, before the 1st of January, 1896, the largest number of yearly subscriptions--over twenty-five--to RANGELEY LAKES, we will make a present of one of those $35 painted boats, made by either C. W. Barrett, H. W. Loomis, or Baker Tufts, as the winning guide may select.

Guides contesting for this prize will bear in mind that all names submitted by them must be those of visiting tourists or sportsmen and not local residents. The proprietors reserve the right to reject any or all names on this count.

The contest is still open, and from now on we hope to see it even livelier than in the past. There are several who are close together, but

The Hustler Will Get There!

The Rangeley Lakes newspaper's subscription contest for guides
(Rangeley Historical Society)

C. M. Barrett is busy these days. He has under way 23 boats at the present time. He will build one that will be placed on exhibition at the Sportsmen's Exposition in New York. Mr. E. S. Osgood, a well known frequenter of the Rangeleys, who is connected with the U.S. Net and Twine Co., Broadway, N.Y., is to place it with their exhibit. As they deal in all kinds of fishing tackle, it is a very fine position for his boat.

The sportsmen's shows of Boston and New York were a yearly pilgrimage for boatbuilders, sporting camp owners, and guides from Rangeley and the other sporting regions of the East. Here, Charlie Barrett could introduce his boats to the public, perhaps take a few orders from men immediately enthralled, and generally talk up the fishing in the Rangeley Lakes.

When the local paper again visited the Barrett shop in late March 1896, it was to make boatbuilding the centerpiece for some diverting trivia. Today, this information provides, at the least, a sense of the work—and tedium—of turning out a large number of wooden boats each season.

(*Rangeley Historical Society*)

Interesting Facts About Barrett's Boats

C. W. Barrett recently patterned or marked out the planking for ten 16-foot Rangeley boats. He has patterns for each strake of planking, there being nine on each side, no two being the same shape; one pattern, of course, being used for both sides of the boat. To pattern each piece a board is selected which is as near the length and curve of the pattern as possible. The pattern is then laid on the board and a lead pencil marked around the pattern at both sides and ends, and the piece is ready to be sawed out.

If the lines required to mark the planking for these ten boats were stretched out in one continuous line they would reach a distance of 6,480 feet or one mile and nearly one-fifth. Six thousand one hundred and twenty feet of these lines must be followed very closely with a saw to saw the strips from the board. This is done with a small circular saw, driven at a speed of about 4,000 revolutions per minute by steam power. To prepare this planking ready for nailing together into the boat, it must be handled over at least twelve distinct times. To nail it together and complete the boat requires on an average 2,718 nails for each boat. Holes for each nail must be made with a brad awl or small drill before the nail is driven. The nails vary in length from ½ to 2½ inches. It also requires 163 wood screws to complete each boat, the screws being from ¾ to 2½ inches long. All of the above lines in marking the 6,480 feet were made with one common round lead pencil—Eagle No. 140.

Much can be learned about the variety of boats and canoes that the Barretts were building in the late nineteenth century from the advertise-

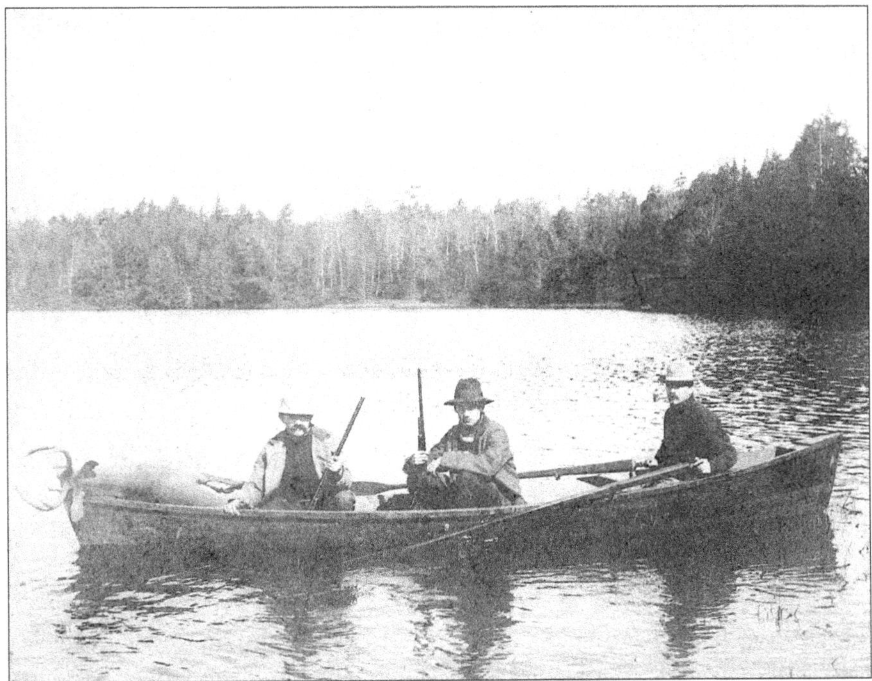

The Rangeley carried hunters as well as anglers *(Rangeley Historical Society)*

ments they occasionally placed in the pages of *Rangeley Lakes*. Barrett ads were announced by the distinctive profile of a Rangeley boat. Below came the information that the shop would build any size or style of cedar boat or canoe, as well as steam and naphtha (petroleum) launches. An advertisement from April 1896 provides further detail. It turns out that the Barretts were generally building Rangeley boats from 12 to 17 feet in length, weighing from 65 to 140 pounds. The price began at $20.00 and ran up to $100.00. The standard size for the Rangeley was 17 feet, a length that, according to Rangeley builders and fanciers, evolved out of functional considerations. At 17 feet, the boat carried nicely between swells on the lakes, making for a level ride. The smallest Rangeleys saw service on ponds and little lakes, where, once again, they suited the environment. And there were means besides length by which form followed function in Rangeley boats. Such was the case with the smooth-sided models called "eggshells." These rare Rangeleys were built for use in the fish-rich Upper Dam pool, where they were thought steadier in the swift water below the dam than their lapstraked counterparts.

The Barretts' cedar canoes were usually 10 to 16 feet in length, weighing 18 to 60 pounds. These were certainly comfortable weights for portaging. Canvas-covered canoes made by the Barretts were both longer and heavier. Normal lengths were 15 to 18 feet and the craft weighed from 60 to 85 pounds. The canvas models were more expensive than the cedar canoes at the low end, starting at $25.00 and increasing to $40.00. The price range for cedar canoes was $18.00 to $40.00.

Besides boats and canoes, the Barrett advertisement mentioned the accessories they also sold. These included the Barrett oarlock, as well as oars and paddles of spruce turned out on the shop's lathe. Edwin Barrett recalls that his family also built fish boxes—containers of pine or spruce with a bottom shaped to follow the hull of the boat. The fish box slid under the middle thwart and held the day's catch. The Barretts likely fashioned fish cars also, again of spruce or pine. A box for holding live fish, the car had a bit of a prow in front for riding through water, hardware cloth for a bottom, and a hinged top. The fish car rode partially submerged behind a Rangeley. When the day's fishing was done, a sport could select those fish to be killed and return the rest to freedom.

In late April 1896 the *Rangeley Lakes* offered an interesting comment on the health of the boatbuilding business in town:

> Baker Tufts has not been able to do any work in the boatbuilding line this season. The other builders have built, and are now building, as many boats as they possibly can. The demand exceeds the supply. Rangeley needs more carpenters and more boat builders. Three or four of the latter could find work at once.

Baker Tufts was an elderly man by this time, and his departure from the scene left only the Barretts and Hod Loomis building boats. There is no doubt that they labored to meet growing demand. As the popularity

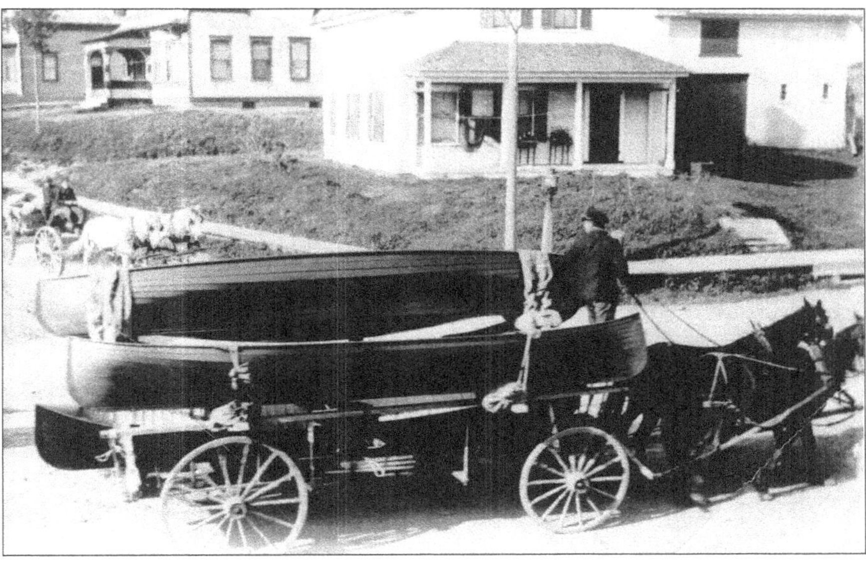

A load of Rangeleys in transit, circa 1900 *(Edwin C. Barrett)*

OARS AND PADDLES ALWAYS IN STOCK

RANGELEY, MAINE. ____ ug.23d.1917, ________ 191

M________ Kenebago Lake Hotel Co,

TO C. W. BARRETT, DR.

BUILDER OF

BARRETT'S RANGELEY BOATS AND NOISELESS OAR LOCKS

TERMS____________ SHIPPED VIA____________

1917				
Aug	3	To 1 square stern boat	55.00	✓
"	6	" 1 spark plug	1.00	✓
		Paid Aug 26: 1917	56.00	✓
		C H Barrett		

of the Rangeley Lakes increased, new sporting camps and hotels started in business and existing ones expanded. Each camp and hotel required a fleet of Rangeleys, and the thirty or so boats that each shop produced yearly was simply not enough.

Along with camps and hotels, there would always be the individual customer. Thus, the May 14, 1896, issue of *Rangeley Lakes* recorded that "C. W. Barrett went to Pleasant Island Wednesday, to varnish Harry Dutton's boats." And one week later, the paper announced, "C. W. Barrett has just received an order for a cedar canoe to be shipped June 5, to T. G. Strong, South Hampton, Long Island, N.Y."

A final glimpse at one boatbuilding shop in the years 1895 and 1896 returns us to the *Rangeley Lakes* subscription contest. In late spring 1896, the newspaper awarded the prize to an ambitious guide who had garnered more than thirty-five new subscriptions. He chose a Rangeley boat made by Charles and Thomas Barrett.

The Barrett shop in Rangeley would continue under its founders until 1928, when Charles sold the business to Thomas's oldest son Frank, and Edwin Barrett joined his brother at the shop. Charles Barrett died before his brother and partner Thomas, who passed away when Edwin was twenty-one.

> He died probably when he was around sixty-eight. We had gone back to Rangeley then, so he passed away in Uncle Charlie's house. He was totally blind. You just can't imagine. Because he had a crippled hand, he never talked much on his hands anyway. To sit there and form letters in his hand and try to tell him what you wanted to tell him was next to impossible. He led a rough life. I felt awful sorry for him.... I wish you could see some of the things he did; well, I can show you some of the things like he made, I can show you miniatures....

A skilled craftsman in the tradition of his father and uncle, Edwin Barrett built wooden snowplows, truck bodies, and cabin cruisers at a shop in Orrington before joining a Brewer paper mill as general repairman. He had chosen to remember his father in the way Thomas Barrett had best expressed himself—with wood. On a shelf in Edwin Barrett's living room were all the items his father made in his Rangeley shop. There was a tiny double-runner sled, a wheelbarrow, a bow and arrow, and more, all perfect in detail. But there was another link in wood, another material legacy, between Thomas Barrett and his son, and it was Edwin Barrett's most cherished possession.

> Well, I was looking forward to when I would get through working up to Rangeley in our shop, and I wanted something to remember the job by, remember what it looked like, so I decided to build this little boat.... This is a miniature Rangeley boat, it's 4 feet long, 10 inches wide in the center, and 4 inches deep. It's a model of the sharp-stern Rangeley boat. It is propelled wholly by oars.... It has oarlocks and oarblocks to control the oars. Everything is made according to the full

boat and it's made to scale.... It has some little round seats in the middle of each seat to keep people in the middle of it when you're rowing the boat. It also has a little floor that you can take out when you want to clean the boat out. It's planked with cedar, it has spruce gunwales, it has oak ribs, and the stem is oak.

Little did Edwin Barrett realize that almost forty years before, his father had built a miniature Rangeley almost identical to his own.

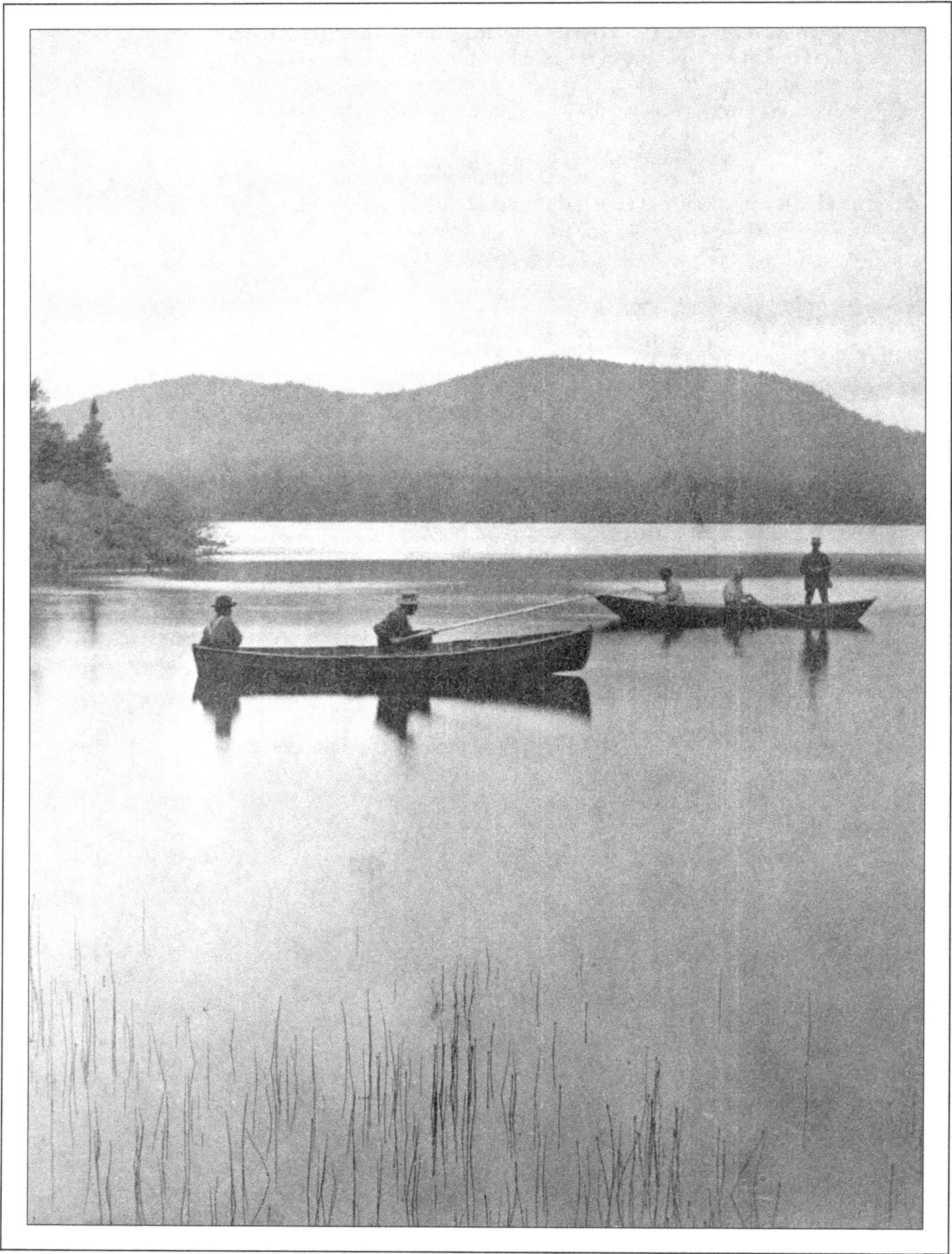

On Kennebago Lake *(Maine Historic Preservation Commission)*

Changing Waters

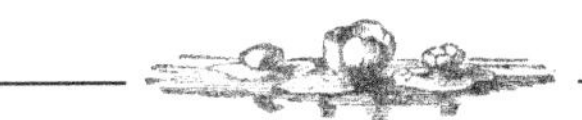

What drew men to the Rangeley Lakes in the mid-nineteenth century was the fishing. Both the size and abundance of brook trout in this primeval, pristine setting were unrivaled. Indeed, the Rangeley boat owes its development solely to the brook trout. But it was not long after this time that men began to alter the very resource that brought them to Rangeley. By the time the Barrett brothers had opened their shop, the biological transformation of the Rangeley Lakes was well underway.

Initially, the lakes contained two coldwater species of note: the brook trout and the blueback trout. The bluebacks were a small trout—never more than 9 inches in length—first observed in the lakes. Their Latin name, *salvelinus oquassa,* commemorated the discovery, as Oquossoc is Rangeley Lake's aboriginal name. Bluebacks spent all but the October spawning season in deep water and served as an important food source of the larger brook trout. Both the brook trout and the bluebacks of Rangeley served as food for humans, too, taken in ways that bore no resemblance to sport. During autumn spawning when they inhabited shallow lake waters and streams, the fish were speared by torchlight and taken in nets. Lumber companies dynamited them to feed woods crews. In accounts of the depredation, the words "bushels" and "cartloads" are often found. Salted or smoked, much of the fish served as provisions during western Maine's interminable winter, but some was sent to market. *Forest and Stream* magazine announced in November 1877 that "The first of the Rangeley bluebacks have come to the market from Maine and will be as usual at Mr. E. G. Blackford's stall in Fulton Market."

Among anglers, the brook trout continued to be revered, and tales grew up, such as that of the "fabled Marble-Morse fish."

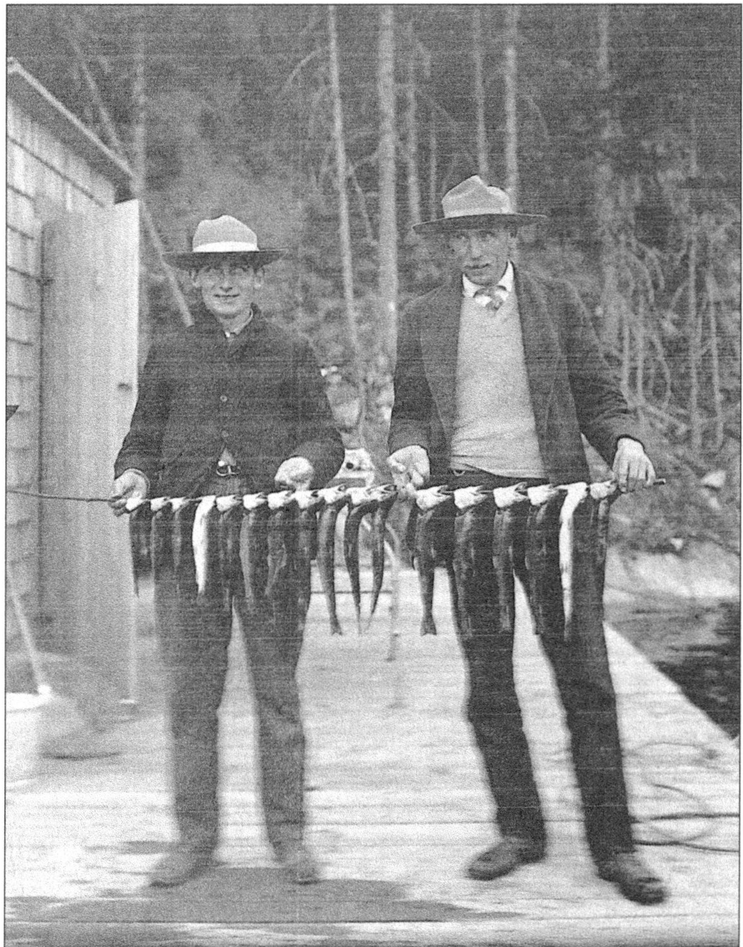

Taking any and all trout *(Rangeley Historical Society)*

He remembered how for several years in the autumn the great trout came alone to the same place in a moderate swirl of water above the dam. He said the fish became the target of many ambitious efforts of both fly casters and bait dabblers. On occasions he would signify his presence by rising to the surface, and with a quiet surface and the sun's rays in a favorable quarter he could be observed lying quietly or slowly moving about. Flies were cast and sunk, also worms, grasshoppers were lowered and trolled in vain, until one day an old guide, who would have scorned to have taken him any other way than fairly, allowed his worm-baited hook to rest upon the bottom for a while, from which it was seized by the fish, which soon lay gasping on the grass.

A fish story based on fact, the trout was taken in September of 1879 by Steve Morse, guide to Mr. Marble. It weighed 11¾ pounds. Until 1914, the Rangeley Lakes held the world's record for largest brook trout caught by an angler, based on a specimen of 12½ pounds.

While fly-casting and trolling were standard fishing methods in the region, so was bait fishing, known locally as plug fishing. In high summer when the trout were deep, a likely spot was chosen and chummed for several days with cut-up minnows. Once big brook trout had been drawn to the area, the fishermen appeared with fresh minnows and, fishing them at great depths, took many large trout. The ever-increasing pressure on the Rangeley Lakes brook trout fishery convinced some sportsmen to enlarge the resource.

With this notion in mind, the Oquossoc Angling Association introduced landlocked salmon to the lakes in 1875, a practice that was occurring elsewhere in Maine, as well. Harland Kidder, fishing guide and later O.A.A. superintendent, recalled the stocking:

> The old Guides Association used to stock the lakes here. Well, my father always maintained that if 10 percent of 'em grew up, you could walk across the water on their backs. They dumped in well over a million fish here a lot of years—salmon. They kept the fishing in good shape.

Almost instantly, the introduction perceptibly changed the ecology of the waters. Most noticeable was the decline of the blueback trout. The diminutive fish now served as food for two predatory species and its population dwindled. By the early 1900s the blueback was extinct in the Rangeley Lakes.

In an attempt to blunt the loss of the blueback trout, smelt were released into the lakes in the 1890s, an activity that Charlie Barrett had a hand in, according to the local newspaper. As the new forage fish for salmon and trout, the smelt served admirably. They also engendered a new spring tradition in town.

> The smelts began running in Kennebago Stream last Thursday night, and much sport among the residents is being enjoyed. "Going smelting tonight?" is the form of greeting heard from young and old, and if you don't reply, "You bet I am," you are no sport, for when the smelts start up Kennebago Stream the whole town turns out to meet them, for they are the first fish of the season, and when salt cod and herring have been the fish course all winter they certainly do taste good. From Indian Rock to a good distance up-stream one can see the glint of lantern and small fires, where groups of men and boys are gathered, dipping smelts from the stream to pails, boxes, and grain sacks.

Smelt served as the first fish of the season for salmon as well, and after May's ice out, the landlockeds fed with a frenzy that produced a like effect in anglers. Spring salmon fishing was always the best of the season. The introducers of salmon had been perceptive in assuming that many more fishermen would discover the region. A correspondent for a sporting journal concluded that by 1883 there were three thousand anglers on the Rangeley Lakes annually.

While salmon thrived in the Rangeley Lakes, their success seemed

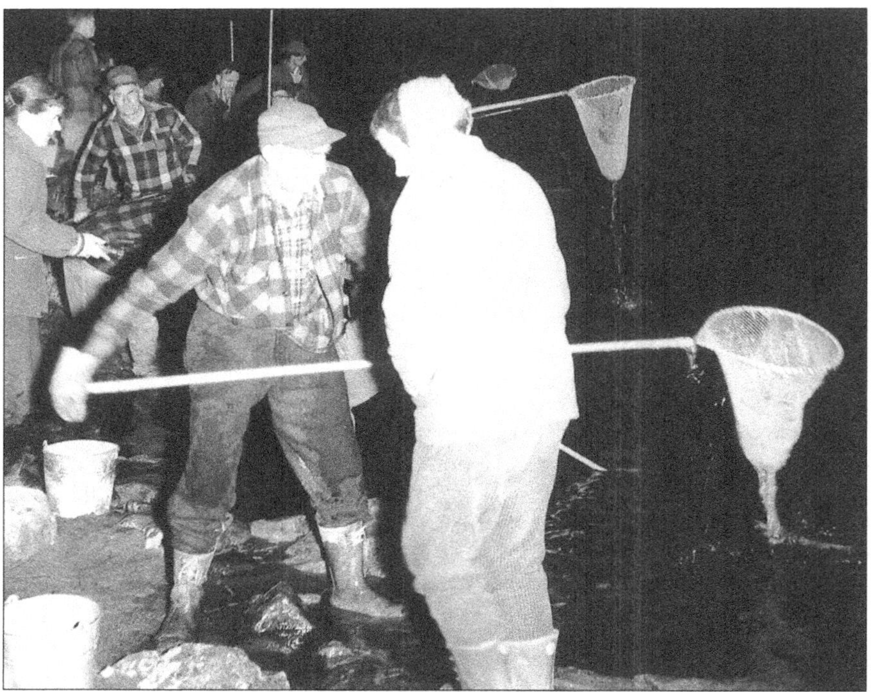

A nineteenth- and twentieth-century tradition—smelting in early spring
(Rangeley Historical Society)

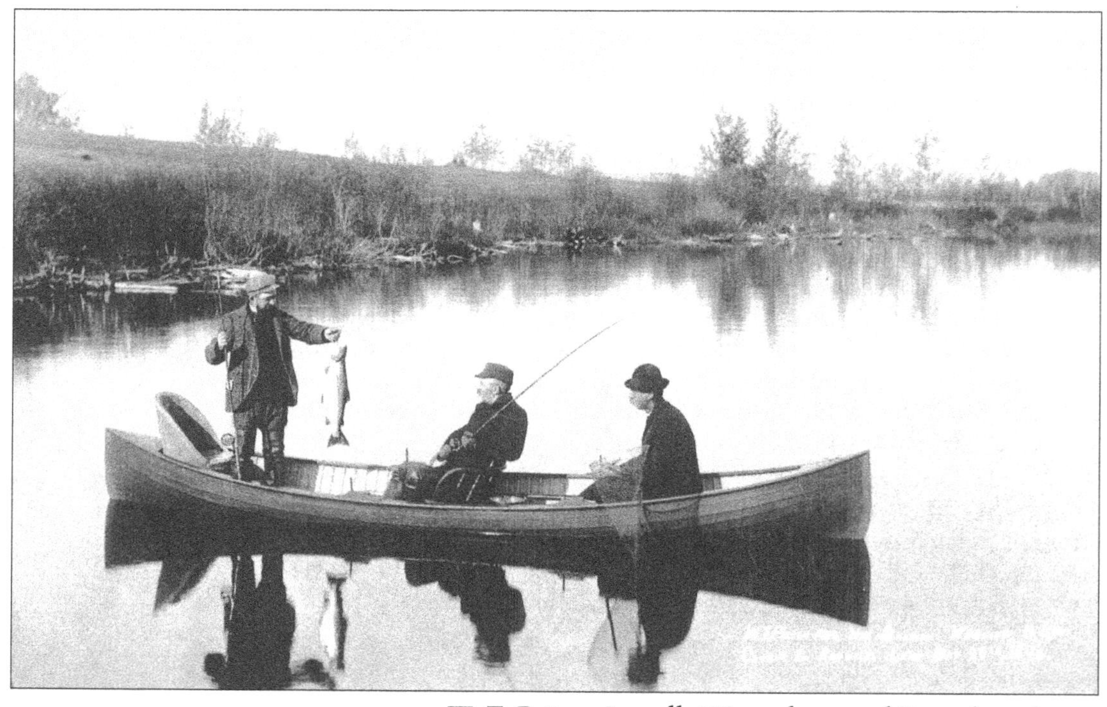

W. T. Patton, Lowell, MA, and a record Rangeley salmon, 10 lbs., 12 oz., May 20, 1896 *(Rangeley Historical Society)*

to come at the expense of the brook trout. In the competition between game fish for food and spawning grounds, brook trout lost out. Despite continued stockings of each fish, records of catches from 1899 forward showed a preponderance of salmon over trout. To some, a game fish was a game fish, but others like fisheries biologist W. C. Kendall mourned the brook trout's decline. "It is, however, a matter of regret to many familiar with the one-time glory of the Rangeley Lakes as trout waters that the salmon was ever introduced. But the evil, if it were an evil, was done, and it cannot be undone." Through human intervention, in twenty-five years the Rangeley Lakes went from being North America's premier brook trout fishery to one dominated by another species.

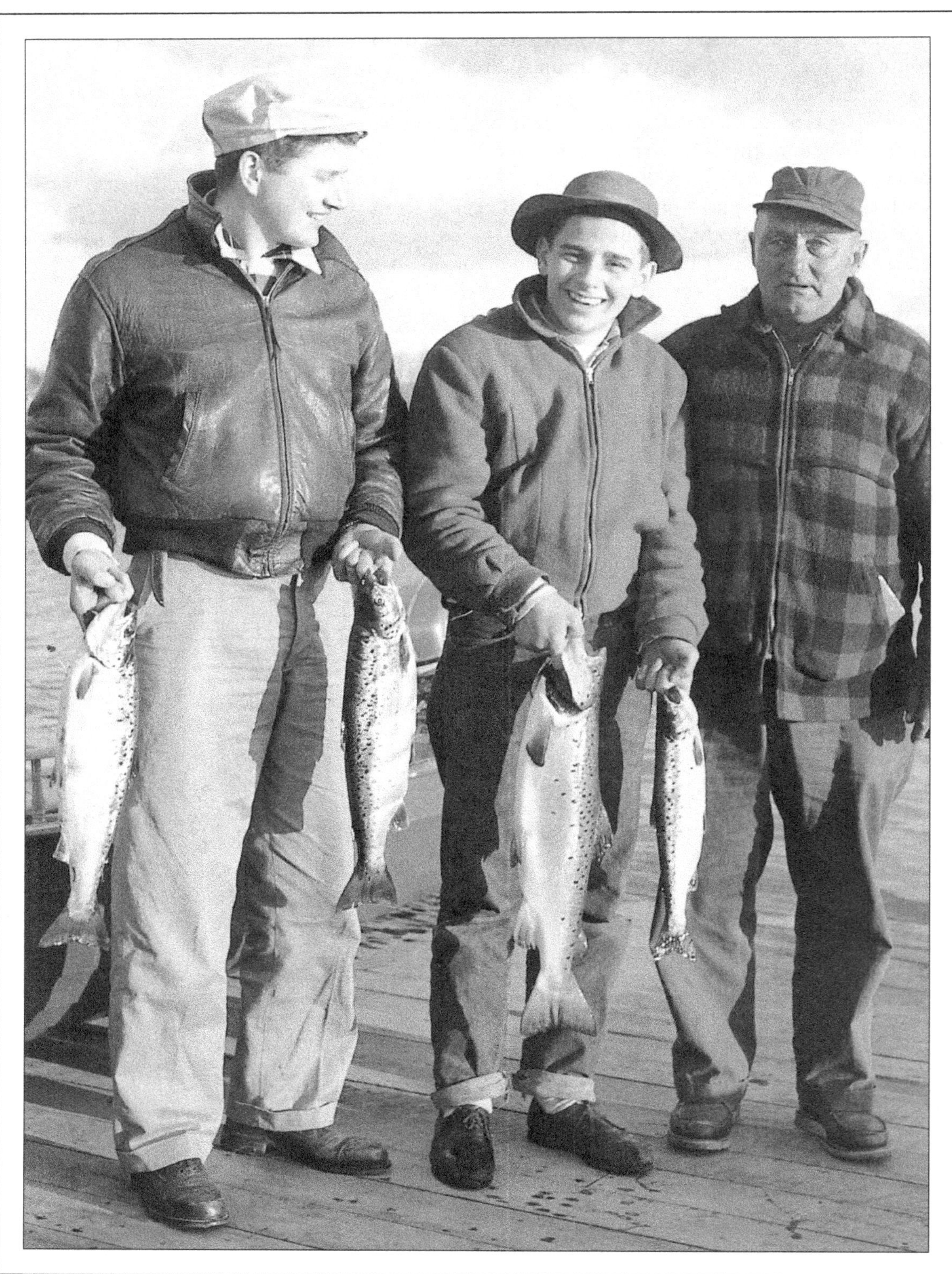

60   **The dominance of landlocked salmon, mid–twentieth century** *(Rangeley Historical Society)*

The Guides and the Camps

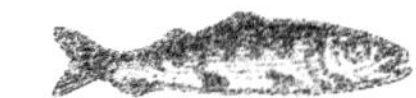

The truest citizens of the Rangeley Lakes—the keepers of the Rangeley boat, finders of fish, and companions of the sport—were the guides. As soon as there were anglers from away in western Maine, there were guides, natives who, whether they were sportsmen themselves or not, knew the landscape and willingly escorted the foreigner around in it. The first and foremost thing each guide knew was that accompanying a gentleman hunting and fishing was an easier way to make a living in Rangeley than by farming or lumbering.

Initially, the guide was attached to a sporting camp, hotel, or private camp. Viewed from the water, the sporting camp complex often appeared tripartite: the woodsy log or shingled lodge commanding a central location flanked by a line of cabins on either side. The location of all these buildings was meant to provide the requisite views and fresh air, yet resolve the dilemma of accommodating the conflicting demands of privacy and community in the same institution. The proximity of the cabins to the lodge ensured easy access for patrons who dined, socialized, conducted business, and met fishing guides at this location. While food preparation and serving rooms were often attached to the lodge, other ancillary structures—outside of the boathouse—were discreetly placed at the rear of the complex. These included lodging for guides, kitchen help, and other employees, and buildings where various commodities such as ice and wood were stored. Essentially, the sporting camp was a self-contained community, like a mill town or Shaker village. Due to the camps' isolated locations, most necessary goods and some services were provided on-site. Nearly every sporting camp boasted in its advertising literature of providing milk, cream, and butter from its own cows, eggs from resident laying hens, vegetables from its farm garden, and local spring water. It boasted, too, of honest Yankee cooking and plenty of

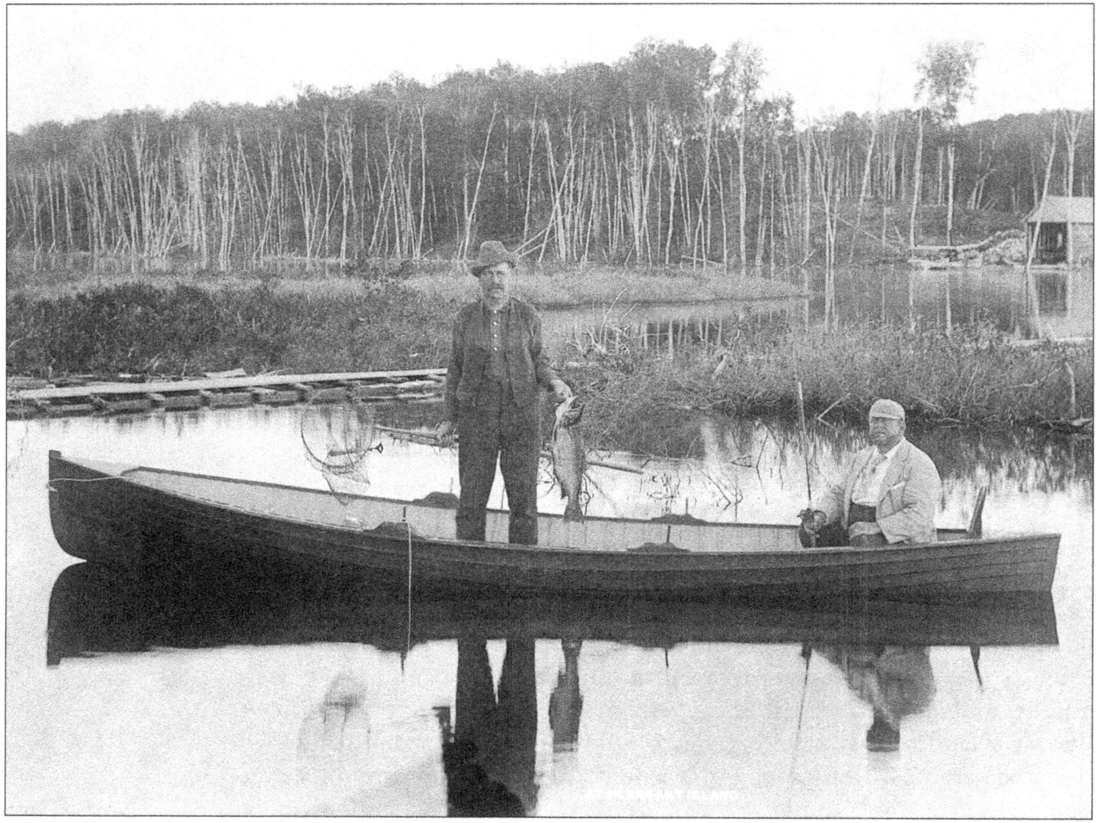

The ubiquitous pose: sport and guide *(Rangeley Historical Society)*

it—oatmeal, homemade bread and pies, roasts, and broiled trout that matched the simple splendor of the place.

When Harland Kidder graduated from high school in 1936, he began guiding at the Barker Hotel, where his father had guided for years. Eventually, he moved to the Oquossoc Angling Association and guided during the strong spring and fall fishing.

> When I started it was five dollars, and that was big money then. That was in the thirties, when the Depression had come back. For the five dollars a day you furnished a boat, but that wasn't all clear money. You boarded yourself. I well remember my father guiding and furnishing a motorboat for eight dollars a day. But in those days, it wasn't a day here and a day somewhere else. You had one party pretty much all summer. From around the first of July to Labor Day, you'd go with the same people and they'd fish every day.

And beyond the daily wage, if the day's fishing had been good, tips were dispensed. Guiding at the O.A.A, where much of the fly-fishing focused on salmon in the Kennebago River, was a highly competitive enterprise, but always bound in decorum.

I well remember when I first started guiding here, it was a horse race out there when the fish were in the river. There were so many guides that you were jockeying for position. Because you never fished in anybody else's water. Anybody that's fishing in the river, they have from their boat downstream as far as their cast is, and a little extra. You just never fished in their water. If you were going up by them, you asked their guide which side he wanted you to go on, and you never fished when you were going by their water. You waited until you got up above where you could anchor. It was just like a horse race, after lunch and in the morning when everybody was going out. All the guides jockeying to get the best spots.

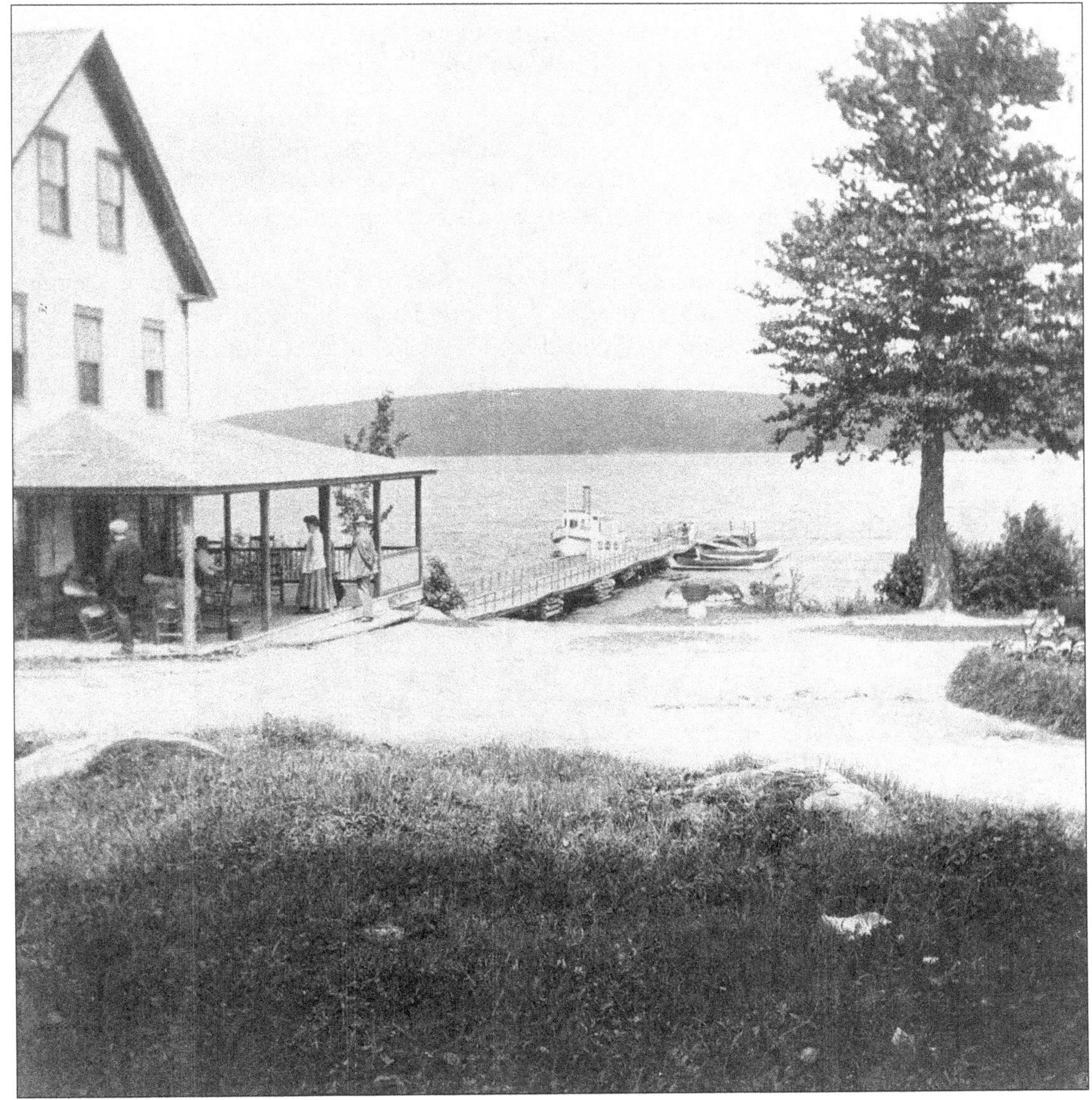

Angler's Retreat, Middle Dam, Lower Richardson Lake, 1909
(Maine Historic Preservation Commission)

In the race for best river position, Harland Kidder made out well. His sport, fisherwoman Juliette Townsend, appeared regularly in the O.A.A.'s angling record book from the 1930s onward. In the association's parlance, "She was high rod here for years." For Juliette, as for all successful anglers, the record book recorded the weight of the trout or salmon, largest fish caught, total fish caught, where the fish were taken, and what flies were used. But there was a distinction at the Oquossoc Angling Association between fish caught and fish killed, for by the 1930s the club was aware of catch-and-release as a means to preserve the fishery.

You could only kill one fish out here a day. I guided her [Juliette Townsend] for five years; I killed one fish for her in five years, and that was a three-and-a-half-pound trout. Her son was going home the next day and he wanted to take it down to a friend.

This emerging ethic puts in context a remarkable entry from the 1930s in the O.A.A. records: "May 6th—170 fish taken, May 7th—fully that number, May 8th—266 fish taken by 17 rods." During three days' fishing in early May, likely at ice-out, over 600 fish were caught, but fewer than 25 were killed.

There was another Rangeley native whose guiding career spanned the profession's two eras. With the demise of the sporting camps and grand hotels around World War II, the guide became a free agent, or in

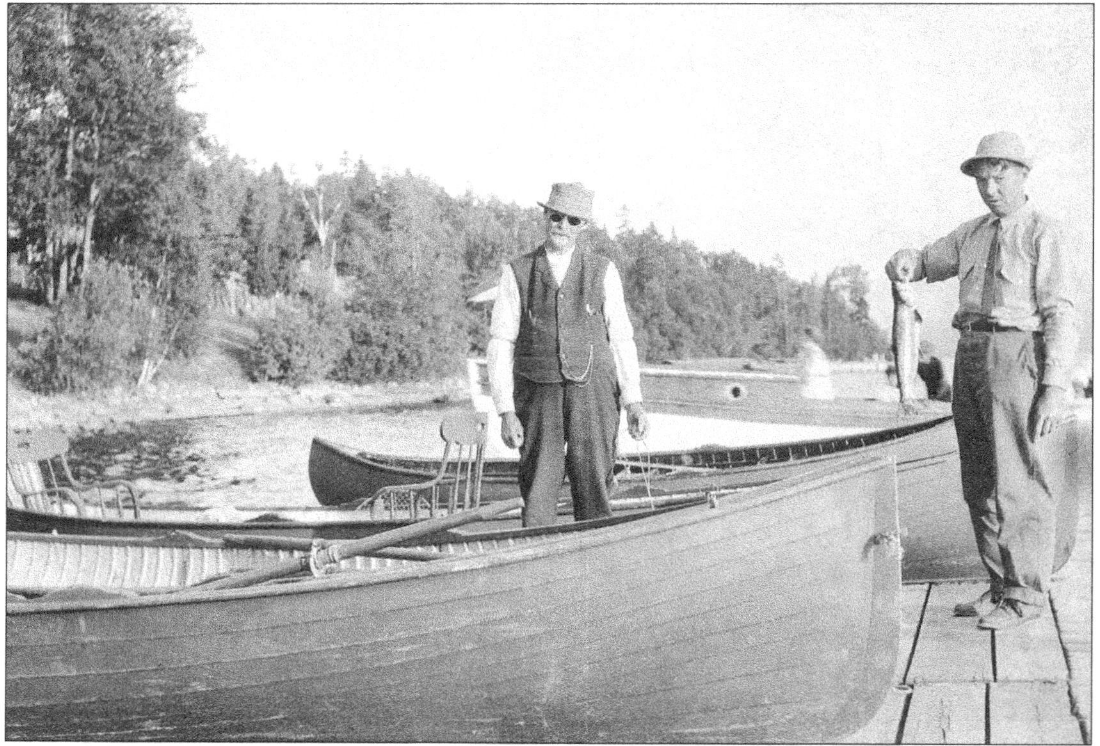

   Which is the sport and which is the guide? *(Rangeley Historical Society)*

our vernacular, a consultant. Walter "Skeet" Davenport began guiding at York's Camps on Kennebago Lake in 1936, and was immediately tested.

> My first trip to Kennebago was one which probably launched me on a career; without real stubbornness within me it probably would have killed my career. The old guides in those days were a lot which never liked to see anybody imposing on their territory. And I remember that first morning I gathered my sport and I walked down to the dock, you know. I was about eighteen years old and on the dock were twenty or more old guides ready to go fishing. And when they see me coming they all turned back-to and as I trooped the line, one man who I knew real well didn't speak to me, but at least he looked over his shoulder as if to say, well, there goes a young worm, or he was saying maybe he'll make it. But that instilled in me a desire to make it.
>
> I went out fishing, got in my boat, rowed over to the mouth of Flatiron Brook, and proceeded to fish with this sport. And we had great fishing. Here I am within sight of the camp, all alone. The other guides, they all got in their boats and rowed out of sight up the lake. So at noontime we came in, we had some nice fish. The guides were gone for the day, apparently. Then at nighttime they came in and they didn't have too many fish, most of them. But we had terrific fishing. So this went on for two days and then J. Lewis York called me in and said, "You know, the guides are complaining." I said, "What's the trouble?" He said, "You're fishing too close to camp." I said, "Well, I thought the idea was to produce fish for the sportsman." He said, "You're right. You keep on fishing anywhere you want to, I've got a real satisfied customer."
>
> By the end of the fourth day, one of the guides came along and wouldn't concede that perhaps I was fishing where the fish were, but he wanted to know what kind of a fly I was using. So I showed him this fly we were using and getting these fish on. In reverse to that, now one of the major places that everybody fishes is the mouth of Flatiron Brook in the spring. People from Grant's get in their boats, tune their motors up in early spring, and go to Flatiron Brook. And I think people on the other end of the lake also do it. But I think that the guides really thought—the old guides—if they fished close to home, perhaps they wouldn't be needed.

When Mr. Davenport passed his initiation into the ranks of guides with flying colors, a guide's wages were from three to five dollars a day. With the Great Depression just loosening its grip on America, this was big pay. The cautious guide saved as much of his wages as possible because he had to winter over on that money. Rangeley was a summer town; beyond lumbering, there was little work to be had come the cold and snow. According to Harland Kidder, who began guiding in the same years as Skeet Davenport, "We always used to say, you make it to smelts and dandelions, you think you got it made."

Walter Davenport continued guiding until the Second World War intervened in his life, and the daily routine during fishing season went something like this:

Well, you get up in the morning—if you're in a camp or hotel, the hotel took care of the feeding of the guest. Sometimes you might like to go out before breakfast; especially early in the spring or even in hot weather sometimes, your best fishing is before breakfast. You usually asked him if he wanted to come in for lunch or to have lunch out. Most of the time you lunched out because depending upon where the wind blew and where you wanted to go in order to get to the good fishing areas, it might take you half a day to get there. And if you had to turn around and row back, why you lost all that fishing time. Take, for instance, Rangeley Lake. Had lunch grounds practically every quarter mile round the shores of the lake. If you'd made a plan to fish the south shore of the lake, why you'd fish along and we used to go up into what we'd call Dr. Stahl's lunch ground. If you got further up the south shore into the Long Point area—a number of camp grounds there—you'd go in and lunch.

The lunch grounds were simple, just an open fireplace and tables, but they became the setting for Skeet Davenport's hearty and well-prepared meals. The young guide had observed that "many of the old guides had certain deficiencies and one of them was cooking."

You went in there and...cut your wood—I got smart and used to take some good dry wood with me—then you wouldn't have to scrounge around and you had good dry wood to cook with. And you cooked up a meal, usually it was either steak or chicken or chops. Most everybody carried steak but I got to the point where I varied and used to have lamb chops, pork chops, chicken, and so forth, and cook them up.

My method of fishing meant that you could go inshore—I always liked to see the fisherman fishing; if your fly or whatever you were fishing with spent more time in the water the more apt you were to catch a fish. Some guides would go in and make a ritual of cooking and they'd cook one thing at a time and take two or three hours to cook a lunch. But I feel, in the method that I cook, you can go into a lunch ground and be back on the lake in an hour and have steak, french fried potatoes, or whatever kind of potatoes you want, a vegetable, and coffee, and, of course, you'd take dessert. We used to take homemade donuts or pie or something. You got back on the water and you spent more time fishing.

I hated to go to a lunch ground and set there for three hours just for the ritual of cooking a few strips of bacon and eating that, then something else, and so on. Most everybody—the old guides—figured that to cook a chop or steak you have to build a fire and let the coals go down. I found out that you could build your fire and after it got done smoking, you could start your cooking of your steak in the front of the fire, standing it upright, your coffee pot going, your potatoes going— all these things were coming together, so that when the steak was done, all the other stuff was done. You put it on a plate, it was hot, and everything you ate together, so it cut down on the time on the shore. As I say, my idea was to do more fishing and less talking.

"More fishing and less talking" meant that the sport and guide returned to camp around 5:00 P.M., and after some final chores, the guide's day sometimes concluded at 7:00 or 8:00 P.M.

> But you spent all day on the water or out fishing. Of course, the thing is if you rowed five hours away from camp you had to come back. And probably when you're coming back, unless you have a favorable wind, you don't row quite so fast as going away.

There was a certain etiquette involved in the care and handling of the guide's fishing machine, the Rangeley rowboat, which went unchanged year to year. For Mr. Davenport, it began in the spring: "You got the boat ready in the spring and you kept it maintained for the whole season…. You always painted it in the spring. You wouldn't paint it a second time unless you had an accident." Barring accidents, the Rangeley went fishing every day of the season. It was from day to day that the care of the boat became critical, and some tenets were strictly adhered to, as Harland Kidder recalled:

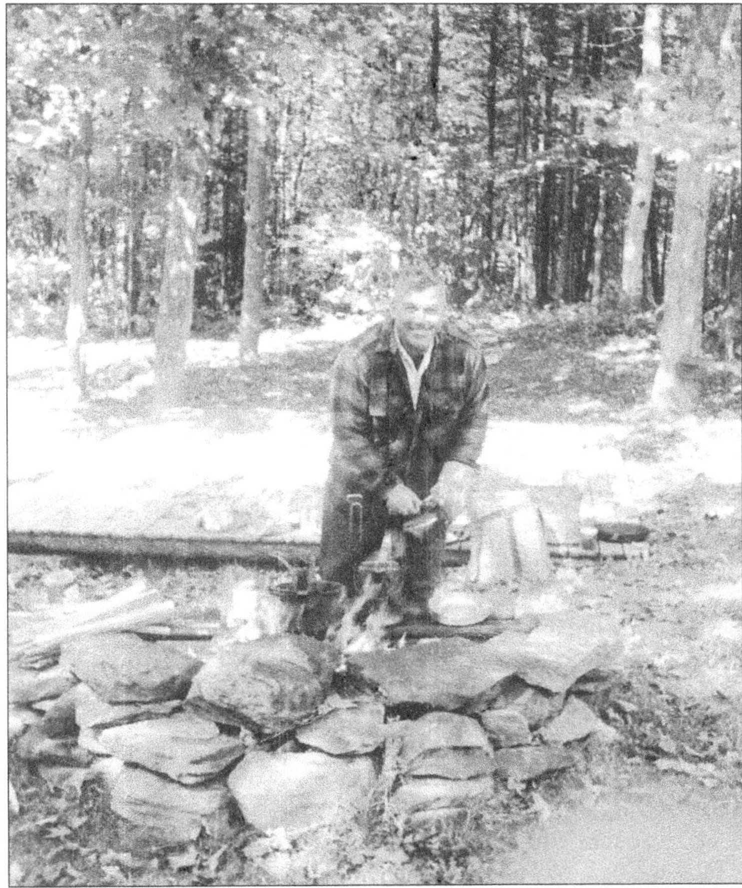

Guide Walter "Skeet" Davenport prepares lunch
(Margaret S. Davenport)

One thing, you never left a Rangeley boat settin' in the water. Always came out. When we came in at noon you pulled your boat up on the float. And now nobody ever pulls one up. They leave them settin' in the water all the time.... You know, a cedar boat, you tie it up beside a float, unless you got some pretty good bumpers, if you got a wind it doesn't take long to chafe quite an area. You can chafe a hole through a piece of planking in just about an hour if it's right. Well, it was just what we always did. I don't know how it ever started or when it ever stopped. We always pulled the boats up.

The Rangeleys not only came out of the water and onto a float, but were cradled, too. Skeet Davenport explained:

In the old days they had these big docks and they were built in an L and inside the L you had a float. And you pulled your boat up onto the float. You used to have a roller so it came up easy. And in the front you had a cradle the boat set into. And it wouldn't roll back and forth. The old Rangeley boats all had plugs in them so that when it rained you didn't have to bail the water out; you went down and pulled the plug and let the water drain out. That was one of your duties, you wanted to make sure that the water was out of the boat before going fishing.... Every now and then you might forget to put the plug in and when you got out it was a little embarrassing. Then once in a while, you might send the kids out without the plug just to have fun with them. It saved tipping the boat up. A lot of the modern boats when they made them they didn't put any plug in them. So you either had to pump 'em out, bail them out, or tip them up on the side. And of course, repeated tipping softens the ribs and you might let them down hard and there's an object there that gouged into the boat, or something. So, it's a feature that ought to be in all boats.

Guides paid particular attention to the condition of the 8-foot oars they pulled on daily, and, as Mr. Kidder made clear, with good reason:

You learned to feather your oars, which you don't see people do too much now. You kept the oars oiled and in good shape.... If they got so they were hard feathering—you know going into the wind if you don't feather your oars you're bucking quite a wind on that oar. And it makes a big difference, you know, at the end of three or four hours' steady rowing.

Beyond feathering, rowing the Rangeley itself required some experience before it became the fluid, unbroken motion of men like Kidder and Davenport, as Skeet reported:

Now, in rowing a Rangeley boat, it isn't like rowing the boats you have in the ocean. Because the oars always overlapped, so you rowed in kind of a rotating motion. You rowed two oars at one time, but the upper hand rotated around the lower hand. Depending on which side the wind is blowing or how you wanted to hold the boat, you'd rotate the oars and pull deeper and harder on one or the other. Of course, the amateur when he started out was always hitting his knuckles and hands, naturally, on it.

Heading out on the water for a day of fishing, the guide and his party positioned themselves according to their number. With a party of one, the guide sat in the middle seat. With two sports in tow, the guide was relegated to the bow seat. Rangeley boats had oarblocks at both locations to accommodate each configuration. Wind was a nearly constant companion on the Rangeley Lakes, and guides took account of this each time they left the float; in Mr. Kidder's memory:

> Well, the guide was in the bow and they [the sports] sat in the middle seat and the stern. You learned very fast how to handle your waves and you're always going downwind. You went out on the lake going downwind, coming back against the wind you'd follow in a little bit closer to shore because the waves weren't as bad, wasn't as hard rowing.

In the boat, the guide became a rowing machine, pulling all day with his eyes on the water—never really looking at the oars. The distances were impressive, especially on a big lake like Rangeley, where Skeet Davenport did much of his guiding.

> Sometimes when you found fish, you rowed over the same ground all the time. A lot of people used to row from here to South Bog, which is about seven miles from one end to the other. Then you had to row back again, so that's fourteen miles. If you got over there and there's a

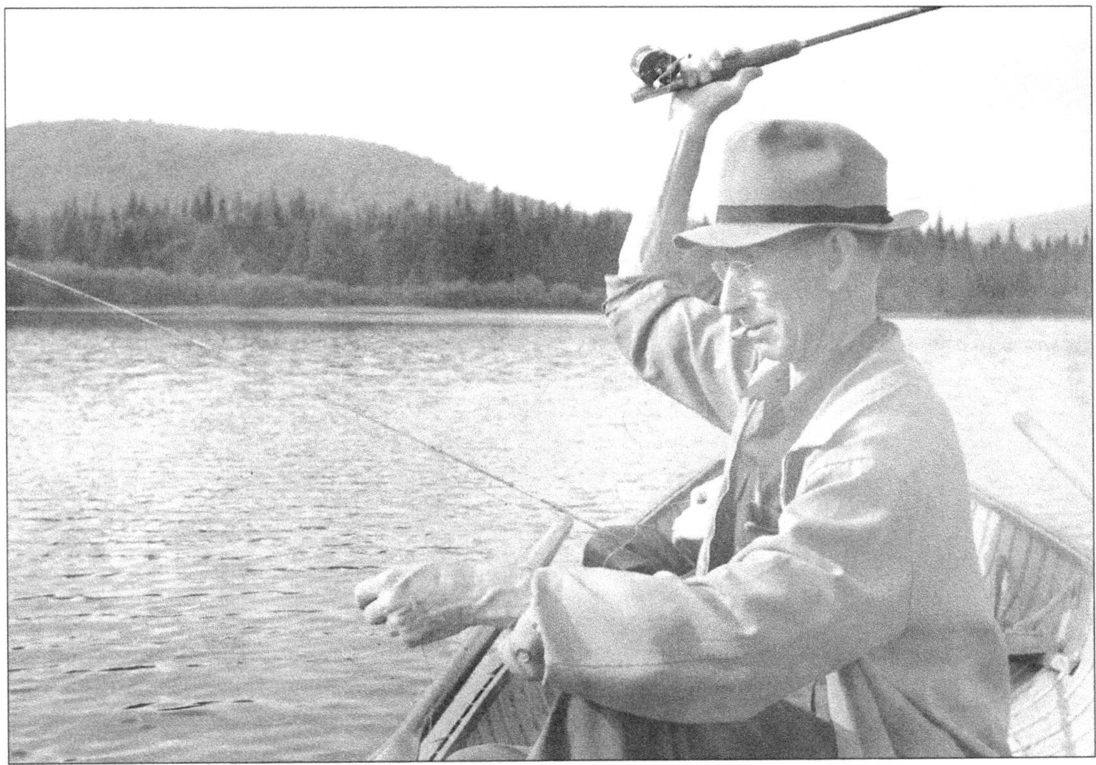

Casting a fly from the seat *(Rangeley Historical Society)*

certain section you want to go up and down twenty times maybe, you'd row up and down. So you put in pretty steady eight hours of rowing and it's going to be sixteen, twenty miles in a day.

The fishing was largely with dry or wet fly, not trolling. As Davenport put it, "You couldn't get a fly-fisherman to drag a fly behind the boat because that was unethical." The sport made his casts from the elevated position of the fly seat, another innovation peculiar to the Rangeley boat. Like the round seat on the thwart, the fly seat is considered another example of guides and boatbuilders joining forces to improve the boat. It was simple enough, a piece of planking perhaps 2 by 12 inches that ran from rail to rail between the middle thwart and the stern seat. A little height made casting easier, and an angler faced the stern—away from the guide—while whipping the water. Fly-casting from a standing position was generally not encouraged, at least not by Mr. Davenport.

Standing up is pretty tough, the boat is a little on the narrow side and also, the smart guide would say, "Well, as long as the wind is down I don't mind." When the wind came up, it would be like rowing a sailboat, 'cause he's sticking up in the air like a sail and it was much harder if you couldn't maneuver. There were times you allowed a fisherman to

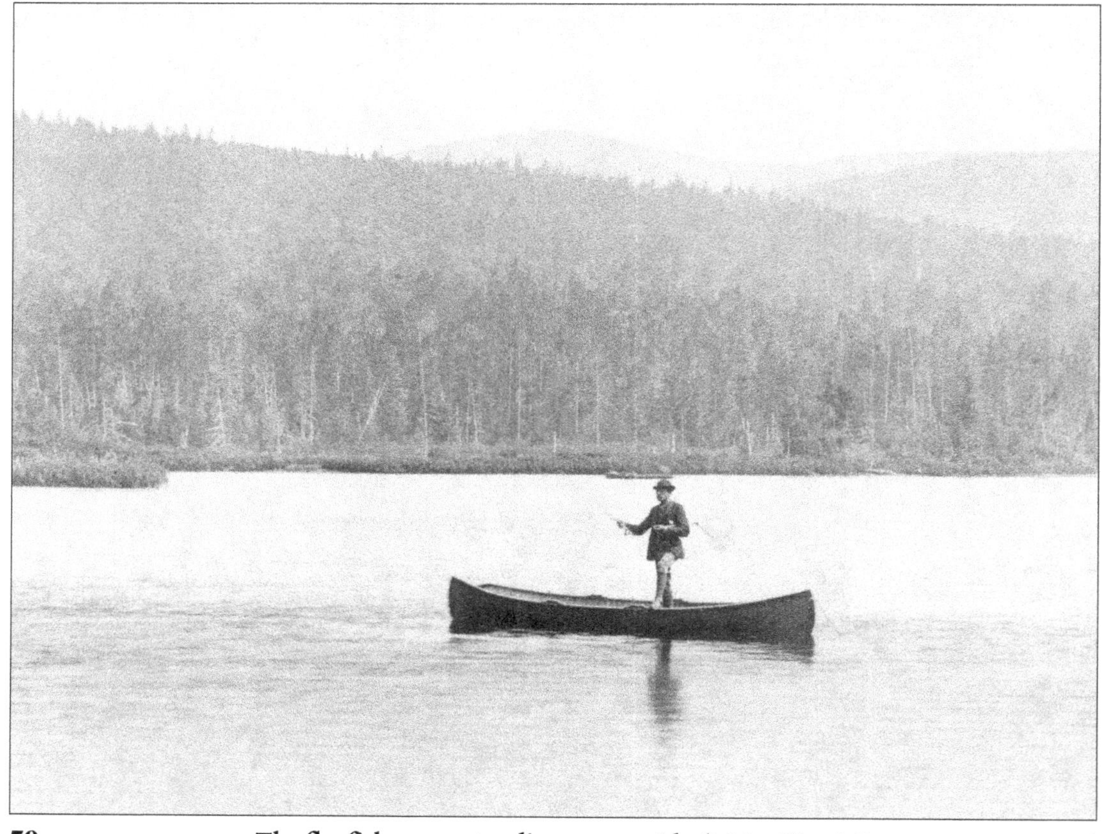

 The fly-fisherman standing, sans guide (*Maine Historic Preservation Commission*)

stand up and the others you didn't. If you were the type of guide that would project your personality to the point where he'd understand, you know.... I've seen some people out there in a desperate wind rowing like the devil and the man standing there like a sail. Well, I was stubborn enough so I wouldn't do that. I felt that you could better maneuver and better fish and certainly didn't take so much out of your hide if the man sat down. And most of the older people, especially, you could hook a chair right onto the fly seat and they could lean back and be very comfortable.

A legless captain's chair could be attached to the fly seat, and a chair with its seat largely cut out was custom-made to fit over the round seats on the thwarts. The captain's chair was found in the St. Lawrence skiff and Rushton's pleasure craft as well, all the boats in which summer people or sportsmen were rowed.

One adaptation only rarely made to the Rangeley rowboat was the addition of a sail. Unlike its parent St. Lawrence skiff, the rudderless Rangeley was seldom fitted with a spritsail and taken downwind. This practice had begun along the St. Lawrence River in the 1870s, when guides would let the air take them home from a day's fishing, and soon was taken up by the summer folks. What lent the St. Lawrence skiff to sailing were the prevailing winds. They followed the run of the river, allowing a guide to row upstream in the morning and sail down come evening. There is no lack of wind on the vast Rangeley Lakes, but it seems less predictable, and perhaps this kept the Rangeley a pulling boat. Nonetheless, some Rangeleys were sailed; boatbuilder Herb Ellis told of coming across double-enders with the telltale block and hole. In fact, he thought that with the addition of a double outrigger, the Rangeley would make a good sailing craft.

In his first years guiding, Walter Davenport worked out of double-ended Rangeleys exclusively, and they were thick on the water. "I've seen on Rangeley Lake right out in front of Nile Brook through the Memorial Day weekend, sixty rowboats out there." The boats Skeet Davenport saw were Barrett boats. "Well, he was the big manufacturer.... I didn't realize until later, when I saw a few of them, that there was any other builder but Barrett. Everybody had their boats built by Barrett." Harland Kidder put it simply, "All your guides had Barrett boats, and that *should* tell you something." What it tells you, these gentlemen agreed, is that Barrett boats were easy rowing, mostly because of a nice sheer. Charlie Barrett's winning personality didn't hurt sales among the guides, either, nor did his claim stated in a newspaper ad, "Remember, I build the best finished boat built in the Rangeley region."

It was during Walter Davenport's and Harland Kidder's years as guides that new technology forced the further evolution of the Rangeley rowboat, an evolution that would conclude with the extinction of the fishing guide. This technology was the outboard motor. In Mr. Davenport's memory, the ensuing developments went as follows:

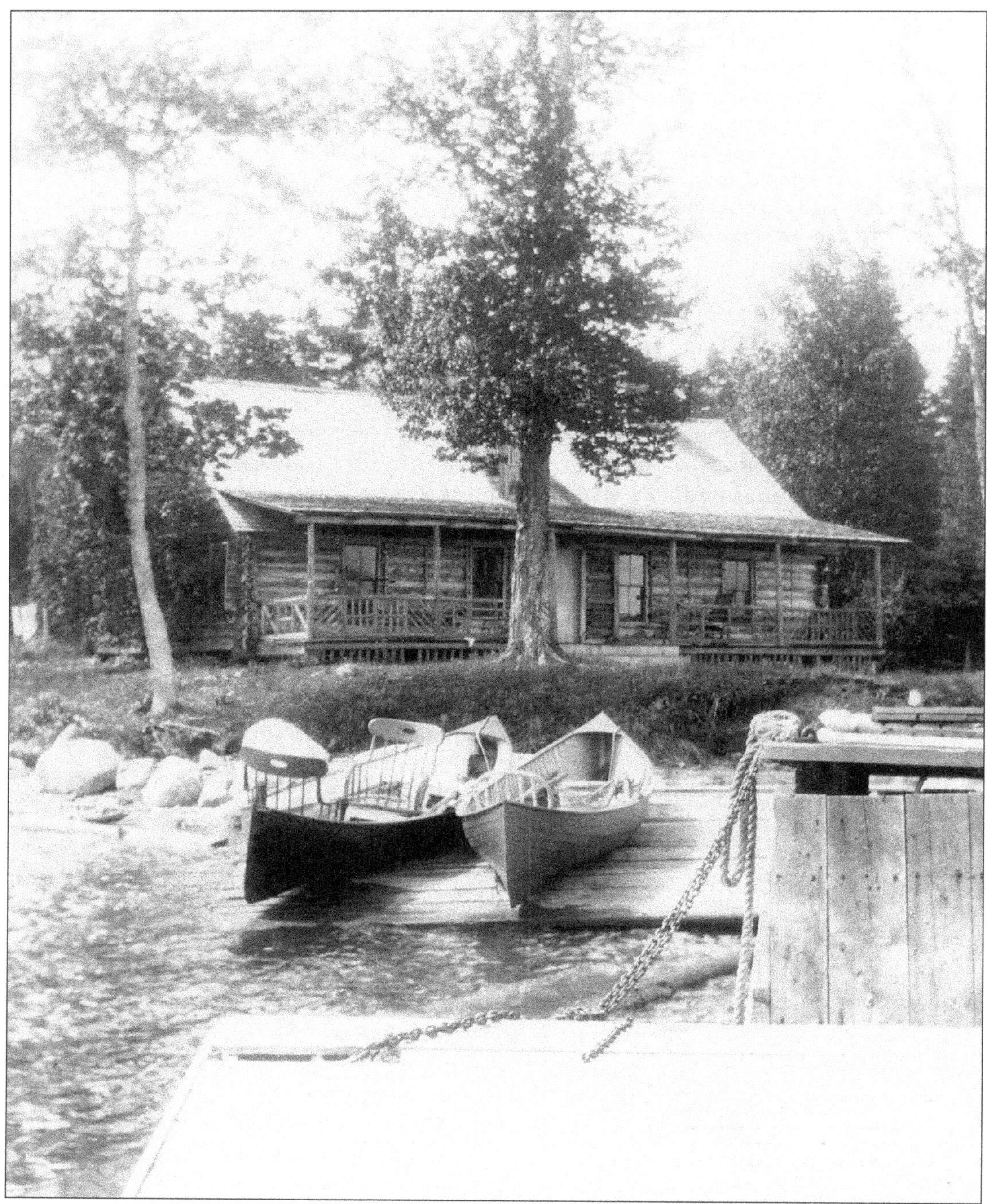

Rangeleys pulled up on the dock, captain's chairs attached
(Maine Historic Preservation Commission)

So they got the motor, they had to find out how to use it. First, I think they put a bracket on the back with a two by four across so you could hook the motor on, but this double-ender was so narrow in the stern, that you could ship water and it wasn't feasible. So then somebody got the idea of cutting a certain portion of the boat off, and they were a little better, but they weren't stable…. These boats which they cut the ends off of them were never really satisfactory because some people cut too much off and others not enough and because the stern was so narrow they didn't maneuver well in rough water. So the boatbuilders saw a need for a boat that would take a motor so they devised the Number One rowboat, or the guide's model.

The guide's model was an elegant compromise between oars and an engine, as it retained the qualities of a capable pulling boat. The problem arose, to Mr. Kidder's mind, when an engine was attached.

But you couldn't slow 'em down enough to troll. I've seen 'em dragging a pail on each side of the boat to slow it down enough so they could use the motor. But we used to use a motor to get to an area if we wanted to and then we always used the oars until outboard motors up in the 1940s got to the point where you could use them in good shape to troll with.

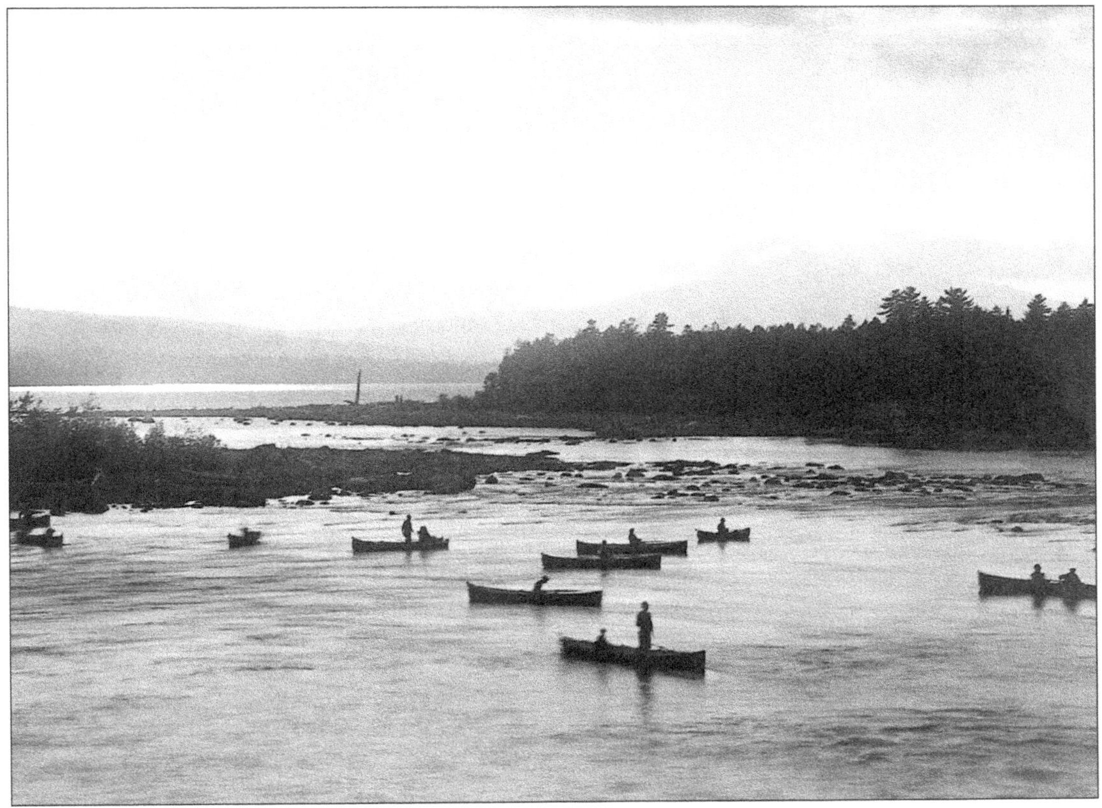

Rangeleys at the Upper Dam Pool *(Rangeley Historical Society)*

In the years to come, builders would continue to flatten and widen the stern section of the Rangeley boat to accommodate larger outboard motors. To Skeet Davenport, it seemed the outboard hastened the departure of the guide.

It used to be that you had to have a guide because most people weren't capable of going out and rowing the boat. If you rowed the boat, you couldn't fish. So they employed the guide to do that. When the outboard motor came along, they had guides initially—a guide was running the outboard motor. Then people began to realize we don't need this guy back here, we can run our own outboard motor and hold a rod out here and put two more friends in the boat.

Fortunately, the tradition of the fishing guide was well established, and the exodus of the sport's companion was not immediate.

In the late 1930s Walter "Skeet" Davenport was offered that most sought-after of all guiding jobs, the position of camp guide. No longer one of many men attached to a hotel or sporting camp, Skeet was now the personal guide of sportsman Frank Smith of Garden City, New York. Under his care, as well, was Smith's camp on Rangeley Lake.

Well, an official camp guide, you had the duties of guiding the person who likes to fish, taking care of the family, you got the camp open, you

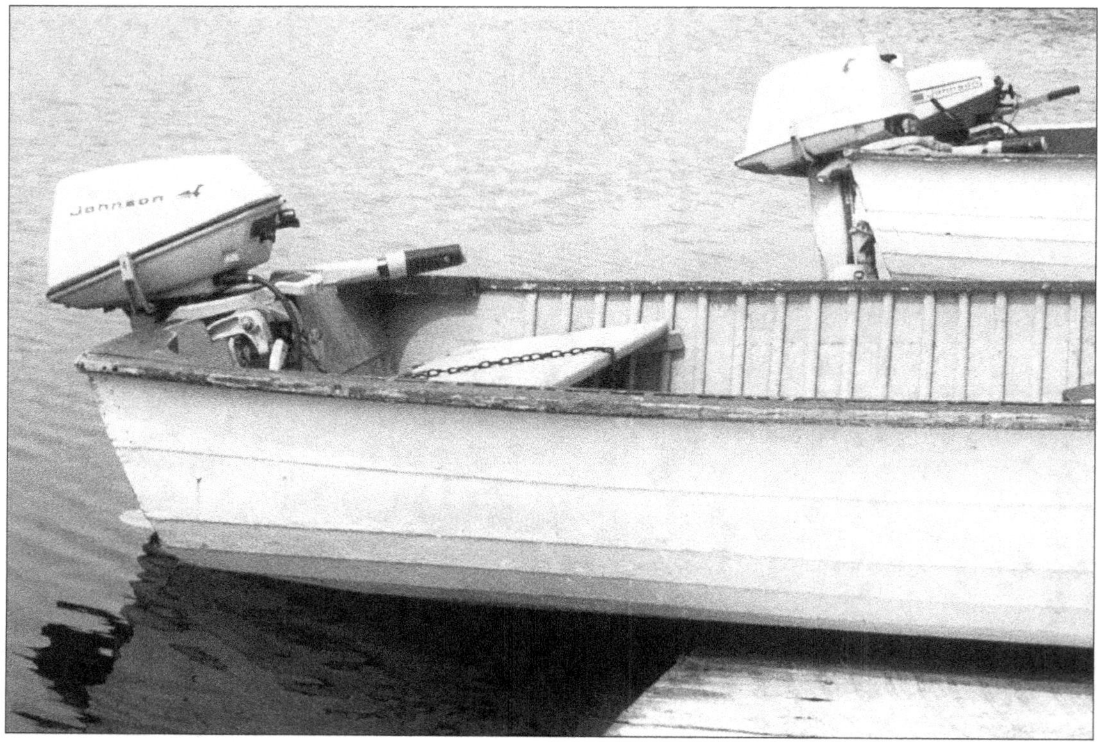

Square-sterned Rangeleys with outboards
(Michael Chaney, Maine Folklife Survey)

closed the camp, you saw that there was wood, and in those days, everybody had ice put up in the winter, so you saw that the ice was put up. You got people in to clean the camp in the spring. In other words, you had other duties which lengthened the season and they gave you a little stipend for shoveling the snow off in the wintertime.

Mr. Davenport remained Smith's camp guide until the 1950s, with time out for war service. In the 82nd Airborne, a notable assault division, Walter Davenport's woods skills were highly prized. He rose to officer rank and saw action in Sicily, Normandy, Holland, and elsewhere. Back on Rangeley Lake after fighting the good fight, Skeet Davenport perfected his guiding skills and formed a close bond with his employer.

You've got to have an interest. I remember two brothers who were quite famous as guides, and one of them, the sports would say he never missed seeing a fish come to the fly. The other brother, he never saw a fish come to the fly, his mind and eyes were somewhere else. He just rowed the boat. The man who really watched that fly and saw the fish maneuvered that boat exactly right so the sport could catch it. That's the difference between two brothers and two guides. So the real good fishing guide was the one who watched the fly.

Important as it was for the guide to accommodate his sport, it was equally important that the sport respect the guide and his skills. Mr. Davenport insisted on such respect; some other guides were not quite so forthright.

I remember being in the boatshop over to Frank Case's, I was getting a motor fixed or something. Frank was getting ready to go fishing in a powerboat with some sports. And all of a sudden, a car pulled up and screeched to a halt, and these people come down and said, "You ready, Frank?" Frank said, "Yep." The man said, "We want to go to South Bog, that's where the fish are, so-and-so caught them over there yesterday." He didn't give the guide a chance to tell 'em where he wanted to go fishing, so they got in the boat and took off for South Bog.... I always felt that if I went to a doctor's office, I wasn't going to tell him what's the trouble with me. I always felt that I knew more about the guiding. Once in a while you'd find a sport who really wanted to guide you, but you either indoctrinated him, or got rid of him sooner or later.

Skeet Davenport shared a warm relationship based on mutual respect with his sport, Mr. Smith. While one was employed by the other, they were companions nonetheless. Each lived a different life and had accumulated varying experiences and skills, which made for good conversation.

With him it was a great rapport. We fished, we hunted, we lunched with the family, and so forth. And I became part of the family, you might say. Most of the clients that I had, I never had many clients that you didn't have a bond...you formed a bond. You still held respect for this person, and you didn't get out of line. You're working for him. I always called my people that I guided by the last name. I always resisted

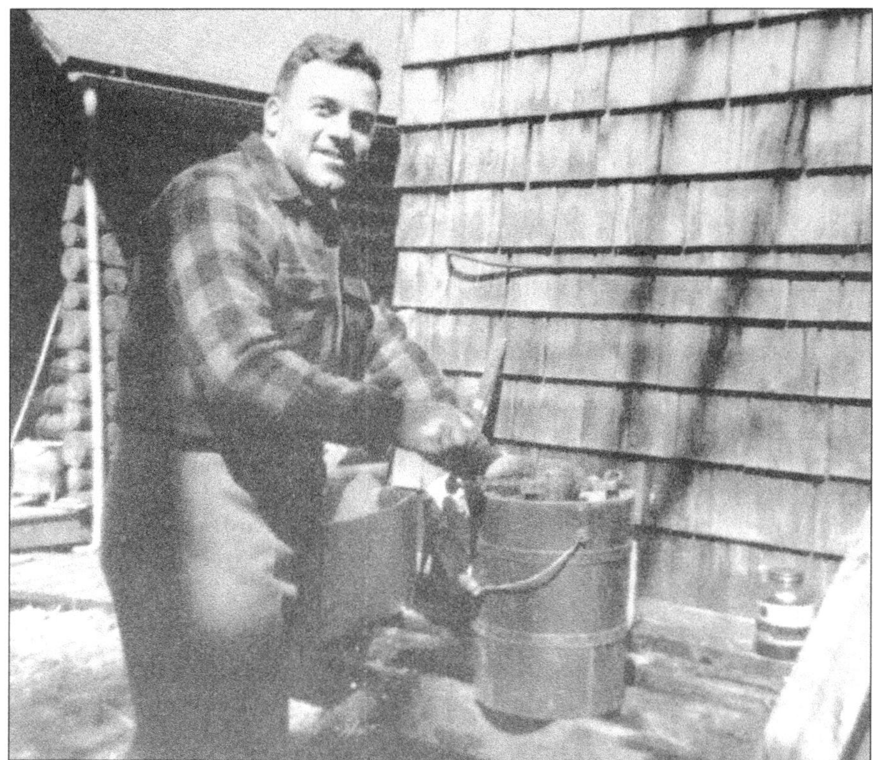

Walter Davenport cranks ice cream at Loon Lodge
(Margaret S. Davenport)

a first-name basis. Of course, now, everyone wants to be on a first-name basis, but I never go on a first-name basis with anyone I'm doing business with. It's just my upbringing.

But guiding was not the stuff of legend, rife with tales of noble friendships between men of character. Like every human endeavor, guiding had another side.

You know, guiding leads into a lot of temptations and one of them is if you guide women. And I found out you could always keep a situation that might get out of control in hand if you never got to a first-name basis with a woman you're guiding. It's just something that I picked up and it keeps things under control.... The second thing is, guides began to drink with the person they fished with. It was easy enough for me because I never drank, still don't. You know, I might take one cocktail or something like that. A lot of times you'd see the guide on the lake and you couldn't tell which was the more inebriated, the sport or the guide. I think that led to the downfall of a lot of guides.

As a new decade—the 1950s—came on, the settled world of the camp guide, the sport, and his family changed. The pace of life was altered and quickened. Rarely now did the sport arrive for late-spring fishing, to be joined on the lake by his family for the languid summer months. The railroad, long the servant of summer people traveling to the Rangeley Lakes, was in decline. Ways in which Americans worked, vacationed, and traveled were transformed and the transformation effected the guiding profession.

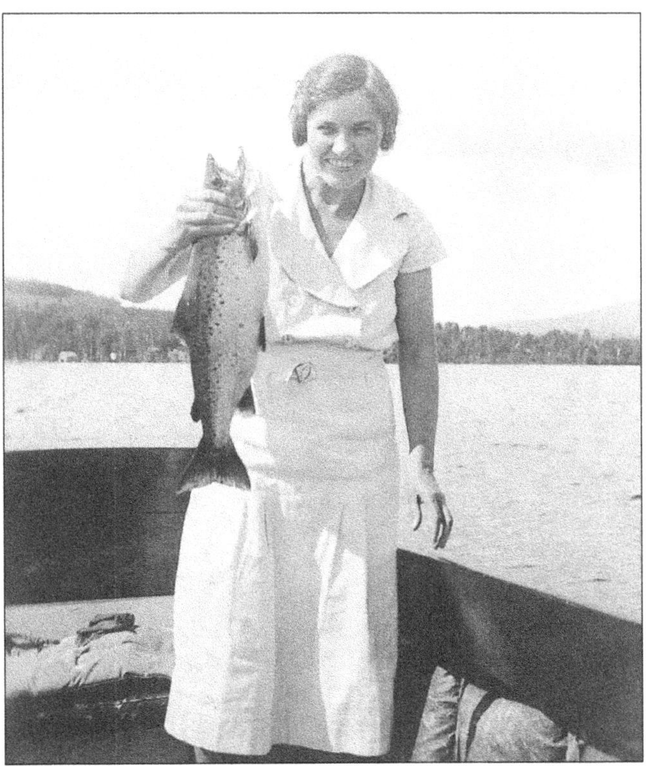

Mrs. Brock and a six-pound landlocked salmon
(Rangeley Historical Society)

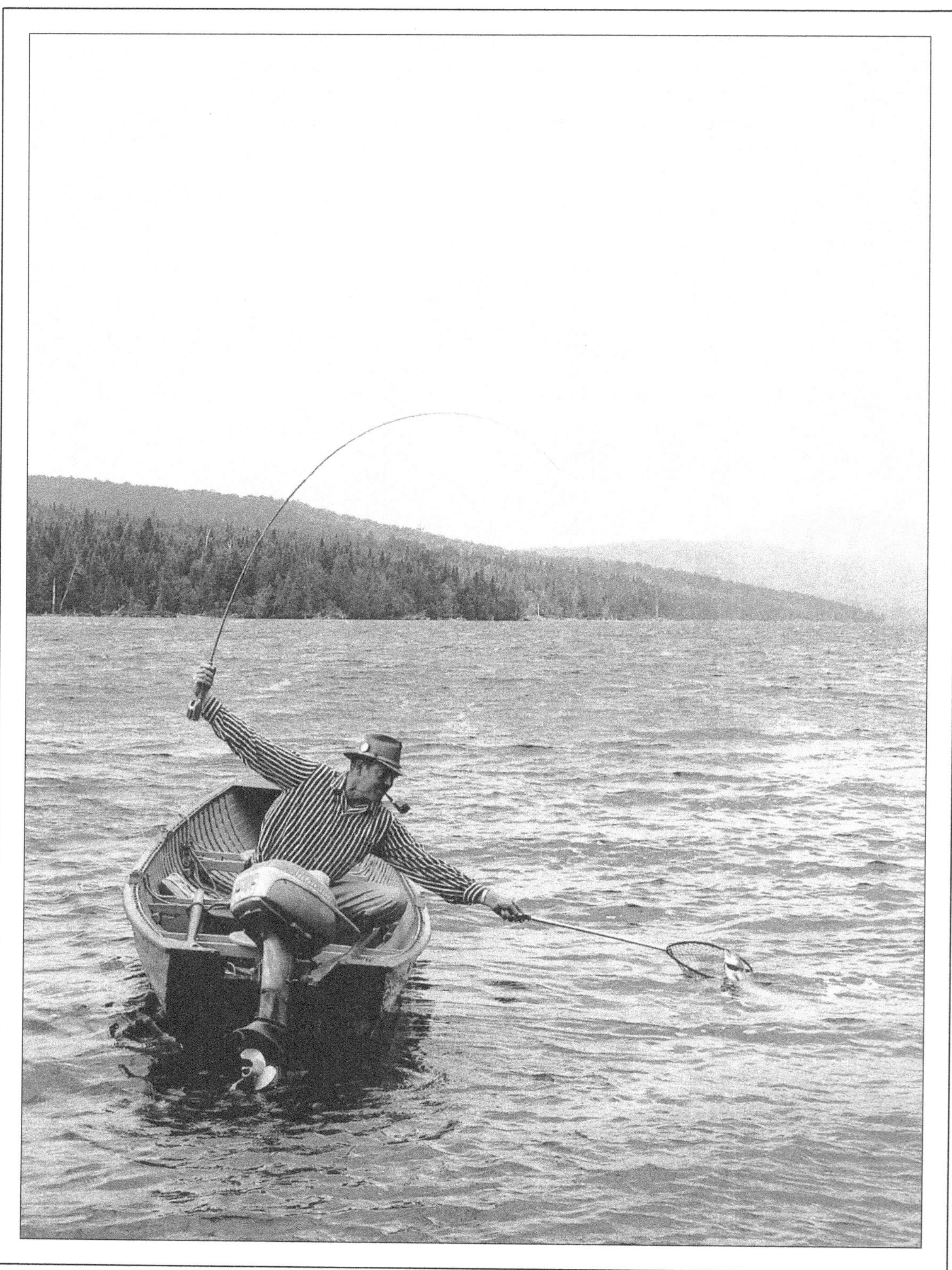

Bringing the fish to net *(Rangeley Historical Society)*

Other Trades

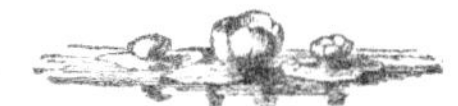

While guiding and boatbuilding were the primary occupations supported by the Rangeley Lakes sportfishing industry, other trades were also nurtured by angling. These involved the capture and display of native trout and salmon—fly tying, rod building, and taxidermy. The practitioners were craftsmen of a high order, and acclaim for their skills spread far beyond the region. Most celebrated of these artisans was fly tier Carrie G. Stevens, who lived for some years near the fishing mecca known as Upper Dam, between Mooselookmeguntic and Upper Richardson lakes. So valuable were Stevens's contributions both to the angling world's awareness of the Rangeley Lakes and the art of dressing flies that Governor Kenneth Curtis proclaimed August 15, 1970, as Carrie Gertrude Stevens Day in Maine. The honor was timely as it came not long before Carrie Stevens passed away.

What seems most remarkable about Stevens's artistry and techniques are that they developed in isolation, outside a tradition of fly tying and without the assistance of experienced tiers. Here is how Carrie Stevens described the origins of her first and most famous streamer fly, the Gray Ghost:

> On the first day of July in 1924 I had the inspiration of dressing a streamer fly with gray wings to imitate a smelt and I left my housework unfinished to develop the new creation. It was a much cruder job than those I have tied since then, but it had two hackle feathers for a wing and an underbody of white bucktail, to which I added several other feathers which I thought enhanced its appearance and its resemblance to a bait fish. Then I felt compelled to try the new fly in the pool and soon was casting with it from one of the aprons of the dam into the fast water. In less than an hour I hooked and landed a six-pound, thirteen-ounce brook trout which I entered in the *Field and Stream* Fishing Contest. The entry

won for me second prize and a beautiful oil painting by Lynn Bogue Hunt, awarded for showing the most sportsmanship in landing a fish. Because the trout was such a nice one and was caught on a new fly I had made, it caused much excitement and resulted in my receiving many orders for flies. Soon I found I was in the fly-tying business.

Mrs. Stevens tied all of her flies hand-held, eschewing the vise, the one fundamental tool of the fly tier. She claimed never to have seen another person tie a fly and didn't allow others to watch her fashion one from beginning to end. This protectiveness of technique was common among many who tied flies. Carrie Stevens's specialty was streamers, the flies tied on long-necked hooks meant to imitate forage fish such as smelt and dace. Tied small, they are cast with a fly rod. Tied larger, streamers

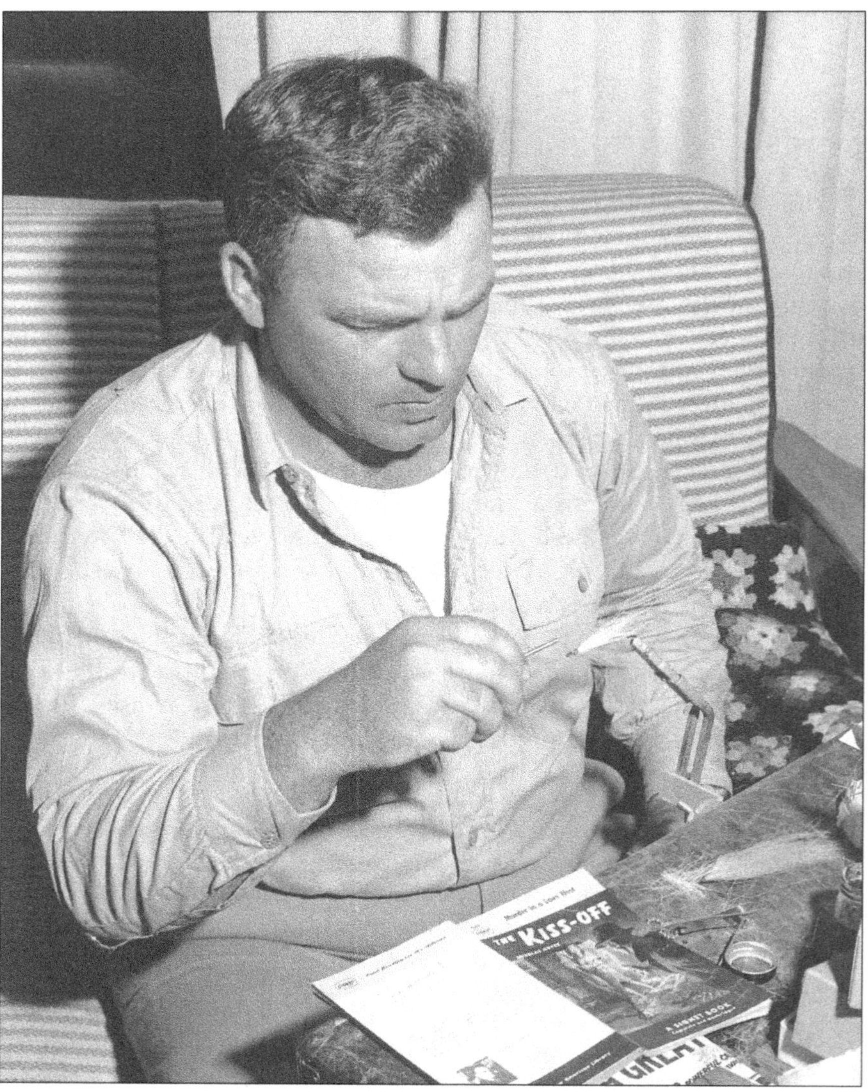

"Skeet" Davenport ties a streamer fly (*Margaret S. Davenport*)

are trolled from a boat and prove especially effective in springtime when smelt are spawning and trout and salmon are active and hungry.

Although Carrie Steven's Gray Ghost remains her most notable fly and is considered one of the ten best streamers of all time, she originated as many as twenty-seven fly patterns. For Frank R. Smith, Skeet Davenport's longtime sport and friend, Carrie Stevens devised two flies: the Frank R. Smith Fancy and the Frank R. Smith Special. According to Mr. Davenport, Frank Smith maintained his faith in these flies year in and year out.

> He used to use two flies—he had two rods—and if he raised a fish on one and the fish refused it, he would swap to the other rod and inevitably, the fish would actually take the second fly. He never fished with any other; in fact, I don't know that he ever knew there was any other kind of fly.

To the discerning eye of fly tier Dick Frost, whose fishing shop sat on the outskirts of Rangeley, several things remain distinctive about Carrie Stevens's work. Consistency is one of them. The feathers and hackle on a Stevens fly are all the same length, fly after fly. She also laid up the materials nicely on the hook, creating both a beautiful and durable imitation of nature. The Stevens trademark appears on each of her works, a band of orange thread tied into the black head of the fly.

If there is a means to explain Carrie Stevens's gift for fly tying, it is hinted at in an advertisement which appeared in *Rangeley Lakes* during 1895. There, Mrs. H. H. Dill called attention to her millinery and fancy goods store located on Rangeley's Main Street. Besides offering worsteds and yarns, hats, ribbons, and embroidery silk, Mrs. Dill also tied trout and salmon flies which she offered for sale. For Mrs. Dill, and then for Carrie Stevens, tying flies grew out of the familiarity with handiwork that was for so long a part of feminine training. A capacity for delicate, intricate work, combined with ingenuity and an artistic bent, made Carrie Stevens an exceptional tier of flies.

Among collectors of sporting antiques, the trout and salmon flies of Carrie Stevens fetch high prices. A single Stevens fly is worth nearly five hundred dollars, while one still attached to the little card that reads "Rangeley Favorite Trout and Salmon Flies" will command even more than that amount. But around the lakes, there are old-timers who use Carrie Stevens's flies for the purpose intended, and that is as it should be.

(Rangeley Historical Society)

During the twentieth century, Rangeley's most renowned man-about-the-lakes was Herbert L. Welch, who arrived as a guide at Haines Landing on Mooselookmeguntic Lake in 1903. Welch's talents were numerous. He was a champion fly-caster, and for a time, was one of only ten members of the Hundred Footers Club—having tossed a fly 124 feet. A born showman, he demonstrated fly-casting at the sporting expositions in major eastern cities, and taught Ted Williams the art. Naturally, he was a guide to the famous who passed through the lakes, squiring around his pupil Ted Williams and President Herbert Hoover. Skeet Davenport remembered seeing Herbie Welch row past with his famous sport on Kennebago Lake. President Hoover was perched on the fly seat of the Rangeley, clad in tie and jacket and wearing a derby, not uncommon apparel for anglers of the day. Not only were Herbert Welch's skills sought after, but also his visage. It is said that his handsome, outdoorsy face found its way into calendars, magazines, and even a Camel cigarette advertisement.

But for many who knew Herbie Welch, his greatest accomplishments came in another field: taxidermy. Welch had apprenticed himself to a taxidermist named Walter Hines in Portland before striking out for the Rangeley Lakes. It is not known when mounting game became common in the region, but it could not have been too much before Welch's arrival. In the 1890s, a sport's memento of his catch still consisted of a profile of the fish cut from birch bark, hand-colored and finished in ink, then mounted on cardboard.

Herbie Welch mounted fish in a little sporting goods shop near Haines Landing that he ran until 1959. A number of qualities distinguished Welch's work from others, not the least of which was that his fish

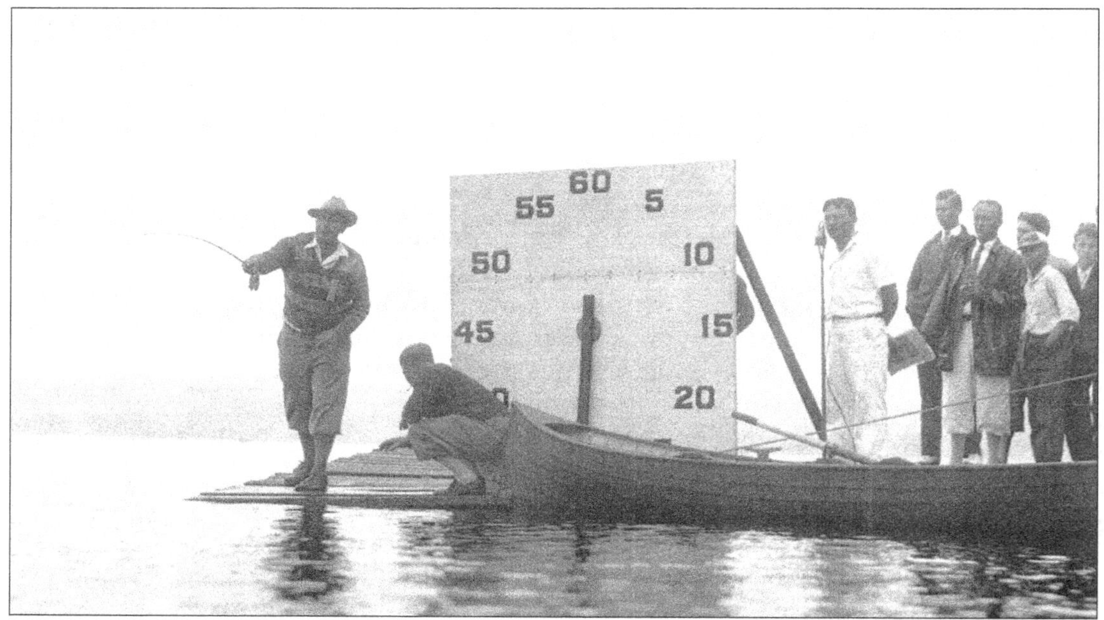

     Herbie Welch competes *(Rangeley Historical Society)*

were extremely well preserved—they lasted. To Walter Davenport, there was a second important characteristic:

> They looked exactly like a fish should. In other words, you look at a Welch fish and it's natural. Too many people want to mount a fish with its mouth open. He never mounted a fish with its mouth open. Everywhere you go in the old camps, you'll find Welch-mounted fish.

Herbie Welch's fish were almost always close-mouthed because, as Skeet Davenport explained, "When you kill a trout or lay it here, it's there with its mouth closed."

Herbert Welch was famous for his mountings of groups of fish. They were often placed against a background of birch bark, and signed WELCH in India ink at the lower left, as any serious artist would sign his work. Perhaps Herbie Welch's most famous effort was that of several trout frying in a skillet, twisting as they sizzled. The taxidermist had captured not only the prey, but the very fate of so many trout.

Not long after Herbie Welch set down roots in Rangeley, a young Nova Scotian named Kenneth Crocker arrived in town and took up the usual work, guiding and trapping. This was Mr. Crocker's life until the mid-1930s, when he decided one winter that he wanted a split-bamboo fly rod. Reading a book on rod building, he went to work, aided mostly by a jackknife. Out of a winter's hobby came a part-time craft, and for Rangeley fishermen, some very handsome bamboo rods. Kenneth Crocker was independently entering a trade that had a tradition in Rangeley. For years, E. T. Hoar had built fishing rods in his Main Street shop; in fact, his business spanned almost exactly the same period as Thomas and Charlie Bar-

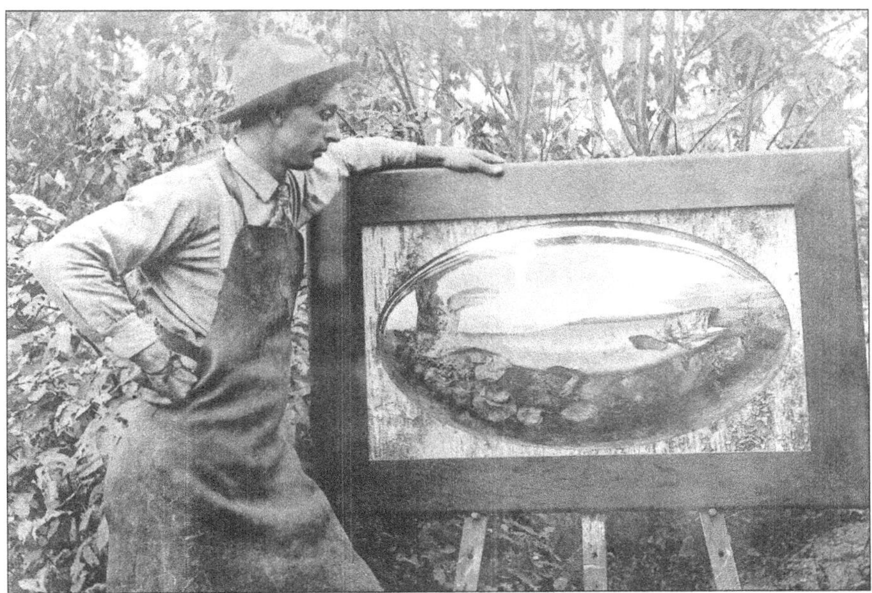

Herbert Welch and a mounted masterpiece, its mouth—unaccountably—open
(Rangeley Historical Society)

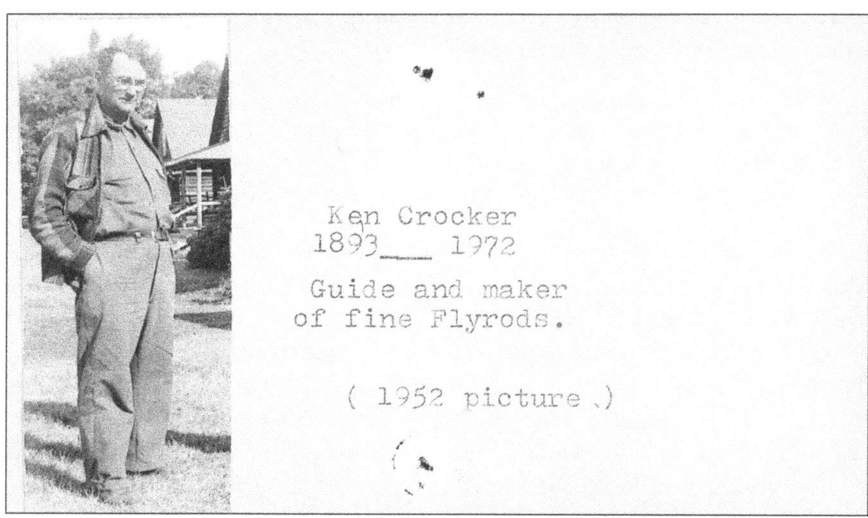

(Rangeley Historical Society)

rett's boatbuilding shop. Edward Hoar built fly and other rods of some exotic woods, as he announced in the *Rangeley Lakes:*

> I make both fly and bait rods, of lance, greenheart, leopard, da gama, bethabara, and hornbeam. Fly rods ranging in weight from 3½ to 8½ ozs. E. T. Hoar, Main Street, Rangeley.

Kenneth Crocker fashioned his salmon and trout rods solely of Tonkin bamboo imported from Southeast Asia, as distant a place from Rangeley, Maine, as any in the world. Then began the transformation of a tropical woody grass into a graceful yet powerful wand. The craft required the simplest of tools—Mr. Crocker did much of the work with a knife blade—but demanded an enormous degree of patience and care. The bamboo was split into thin sections, six for each fly rod. Colored with the flame of a blow torch, the bamboo sections were then handplaned and glued together to form the rod.

Then came the part that was hardest on Kenneth Crocker's children. Each newly glued rod was bound in string along its whole length, rolled under their father's palms on a tabletop to make it true, then suspended from the living room archway for a long, slow drying. All the while these cocoons hung in the house, the Crocker children were forced to tiptoe around them, and woe to the child who disturbed one.

Forming a handle of cork sections, aligning and attaching the ferrules and rod seat, and varnishing nearly completed a Crocker rod. But here was where Mrs. Leath Cameron, Kenneth Crocker's daughter, made a contribution to her father's craft. Mrs. Cameron sewed the cloth bags in which Crocker fly rods were sold, and carefully wrote the maker's name and the rod number on each completed piece. Crocker's earliest rods sold for $50.00, and no Crocker fly rod ever put the purchaser back more than $125.00. All came with Kenneth Crocker's explicit advice on the kind of reel to purchase and the line weight to use. That a guide and woodsman should have turned out lovely fishing rods on the Rangeley Lakes comes as no surprise, but what made Kenneth Crocker's skill inexplicable to Skeet Davenport and others was that it seemed out of character.

> He made a real fine, elegant rod.... Most anybody who done much fishing had one or two Crocker rods, that is, the old fishermen.... When you looked at him you wouldn't think that he was a rod builder. He was a nervous type and couldn't set still, just not the type that you'd think would sit down and do a rod. But when he built a rod, it was perfect.

Kenneth Crocker's trout and salmon rods were, in their day, considered comparable to the bamboo fly rods of Bangor's F. E. Thomas Company and to those of Orvis in Vermont. Although he has been dead for nearly forty years, Mr. Crocker's rods are well remembered by fishermen who have used them. In Rangeley, his daughter Leath Cameron long received calls from anglers hoping to learn where a Crocker rod could be found. There is nothing like the passage of time and the perspective it provides to help us determine the importance of things. Some years ago, the Maine State Museum mounted a major exhibit titled "Made In Maine." Still to be seen in Augusta, the exhibit showcases outstanding examples of manufacturing technology and craftsmanship of the nineteenth and early twentieth centuries from around the state. For "Made In Maine," the curators drew from their collections a Golden Witch streamer fly and a Gray Ghost streamer fly, both tied by Carrie G. Stevens of Upper Dam. Also in the exhibit is an artfully mounted landlocked salmon signed Welch. And suspended above the visitor's head hangs a forest green Rangeley boat attributed to the Barretts. If the Maine State Museum owned a Crocker fly rod, it, too, would doubtless be displayed.

These boats are planked with the best State of Maine Cedar 5|16 inch thick, and are ribbed with oak ribs spaced 2¾ inches apart at centers.

They are of lap streak construction with white lead between laps. They are light but strongly constructed to withstand years of hard use.

Each boat is equipped with sockets to row from either bow or middle seat and one pair of oars equipped with Barret Rangeley oarlocks.

These boats are 17 ft. long 48 inch beam and 16 inches deep.

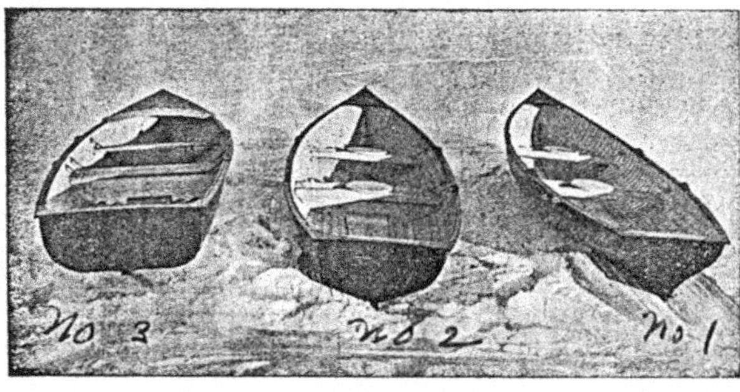

Price No. 1 . . . . No. 2 . . . . No. 3 . . . . . .

Prices subject to change without notice.

The Rangeley Boat in a New Century

With the twentieth century, a new order of boatbuilders arose in Rangeley, gradually taking up the trade that for years had been the dominion of Tufts and Loomis and Barrett. It was a good time to enter the business, as new hotels were built in town and thrived. Altogether there were at least fourteen sporting camps and hotels in operation by 1909. While hotel guests were not the dedicated sportsmen who lodged at Rangeley's sporting camps, they did enjoy a little fishing or a row on the lake, and the hotels all maintained a fleet of Rangeleys for such diversion.

In 1909 only Charles and Thomas Barrett remained of the old guard. They were now joined in boatbuilding by Fred Conant, Arthur Arnburg, and E. L. Haley, who constructed launches. In the years to come, Rufus Crosby, the team of George Bridgham and "Fin" Tracy, Harold Ferguson, Harold Fuller, Martin Fuller, Frank Barrett, S. A. Collins & Son, and Herbert Ellis would all make a living building Rangeley boats. Indeed, this listing of builders may only scratch the surface. Every Rangeley enthusiast who knows the lakes seems able to add one or two to the number. One hears about a man named LaRochelle, who built on Kennebago Lake, and of Roland Ripley, working in the Magalloway region. And there were numerous guides who spent a winter in a boatbuilder's shop making a craft of their own.

The ways in which new men entered the trade varied. It was common for some who had worked in the shop of a master boatbuilder to simply buy the business—patterns, tools, stock, and building—and carry on the work. This is how it happened for Herb Ellis, the last in the direct line of Rangeley builders. In 1938 he purchased the Barrett shop from his employer Frank Barrett. Charlie's nephew and Thomas's son, Frank had taken over the family business in 1928 and run it until he was too ill to continue.[1]

---

1. For a closer look at Rangeley boatbuilder Herbert Ellis, see Paul McGuire's "The Rangeley Tradition," in *WoodenBoat* No. 39.

This practice of the trade being handed down from employer to employee was not always so clear-cut. Fred Conant worked for old-time builder Hod Loomis, and the story goes that on leaving work each evening, Conant took with him another Loomis pattern on paper. Conant eventually built boats on his own.

Some Rangeley builders seemed to materialize from thin air, outside the established lineage of the trade. For example, Arthur Arnburg had clerked at a hotel for years before turning to boatbuilding. He appears to have worked up his own patterns and for a span of time, turned out the most graceful Rangeleys on the lakes. Fishing guides prized the Arnburg boat even above the Barrett, for it was that much leaner, with even better sheer and flare, and rowed like a dream. Seen bow-on, the exquisite lines of an Arnburg Rangeley recall a Viking boat.

Woodworkers wanting to build Rangeley boats had one more means of entering the business. It was simply to buy the patterns of an established shop and start in. Herb Ellis's tale about H. M. Ferguson's beginnings points up this fact.

H. M. Ferguson, who I knew well, he bought the Rufus Crosby. He was telling that all the guides said if you are going to start get the Rufus Crosby, it's the best boat on the lake. He said I bought it and then the bastards wouldn't buy one. He said he had to go and buy Arnburg in order to sell a boat and he bought Arnburg out in '23.

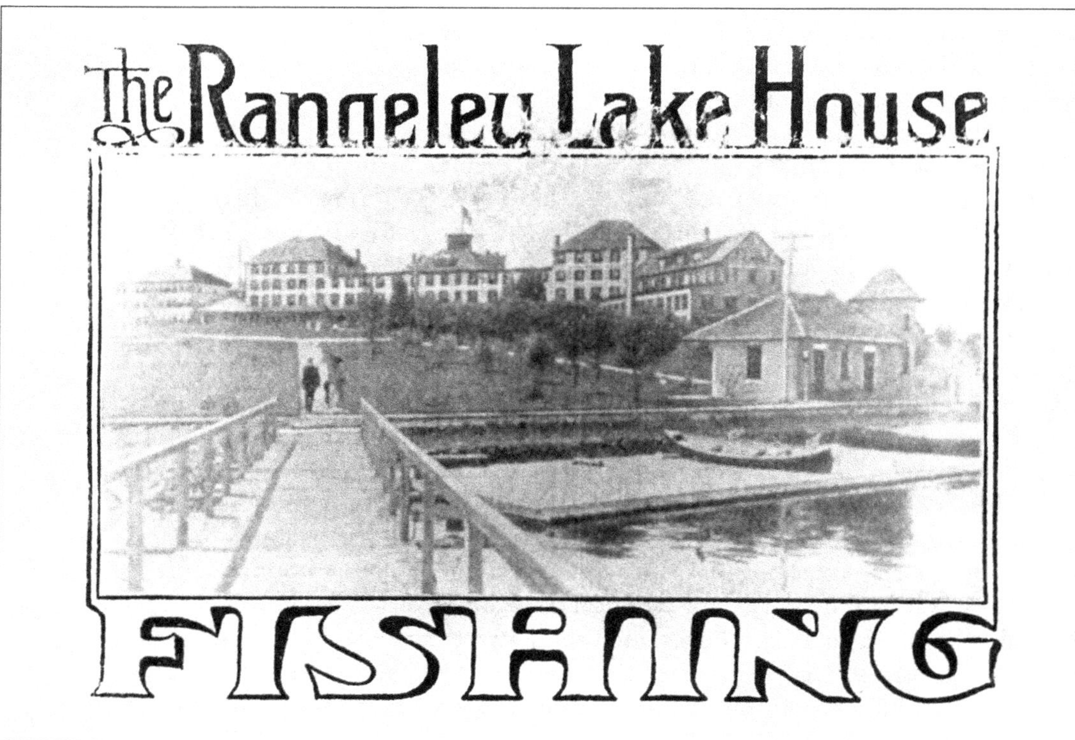

(Apprenticeshop of Atlantic Challenge)

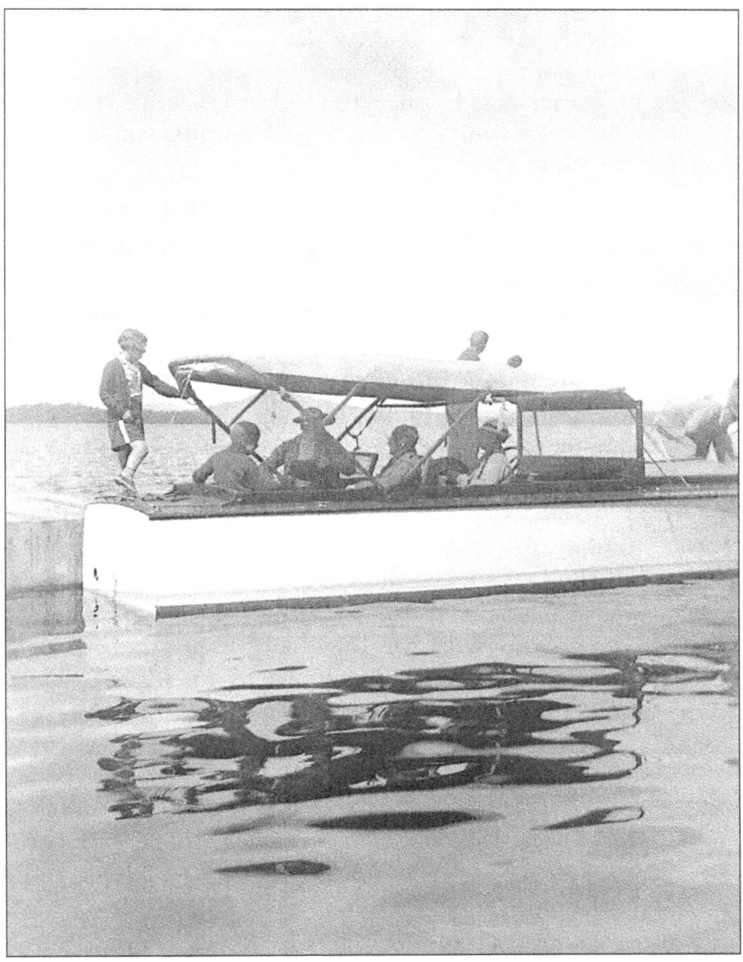

The Gimbel family's cabin cruiser built by E. L. Haley
(Rangeley Historical Society)

Since at least 1908 there had been a boatbuilder of another sort working in town along with the makers of Rangeleys. His name was Ernest L. Haley, and until his death in 1944, he produced small motor launches and large cabin cruisers for well-heeled customers. Haley arrived in Rangeley to work as a house carpenter, and after a while saw a need for touring and trolling boats going unfilled. He decided to fill that niche, and in his time built 130 craft, many of mahogany and comparable to the elegant Chris-Crafts. Ernest Haley was not lacking in entrepreneurial skills, either. He also served as manager of the Rangeley Motorboat Club, and undoubtedly encouraged the use of motor-era craft on the lakes. Walter Davenport recalled the heyday of these boats as follows:

> This was a rich world and a lot of the big camps had big powerboats—Chris-Crafts and different things. The old-time Chris-Crafts, they were a nice fishing boat if you wanted to troll from it. They weren't

any good for fly-fishing, but for trolling or just for comfortable excursions, everybody had them. In those days, we had a big boathouse that was on the edge of the lake here [Rangeley Lake] and probably that was 300 feet long, each boat having a bay and that was full of these expensive boats that these camps owned. Pleasure boats, you might say, but they did fish out of them.

If one were to think that E. L. Haley was no more than a gifted woodworker, a single visit to the Rangeley Lakes Region Historical Society dispels the notion. Haley was interested in more than boats and motors, as his collection there points up. The collection is of eggs, as Haley was also an oologist, a collector and student of bird eggs. Haley acquired native eggs, from the loon's to the hummingbird's, and traded some to distant oologists for exotic specimens. In the thickly padded drawers of a large wooden cabinet the eggs sit. Ernest Haley's birds' eggs may one day be more widely known that his boats, for the Smithsonian Institution has offered—unsuccessfully to date—to purchase the collection.

With a number of new builders in the trade and the advent of the outboard motor, the first half of the 1900s saw some variations develop on the theme of the Rangeley boat. The guide's model or Number One transom had come about with the first truly usable outboards of the 1930s, and later, the wider Number Two transom was added to accommodate motors with as much as seven horsepower. The still-wider, Number-Three-transom Rangeley was largely built in the years after World War II,

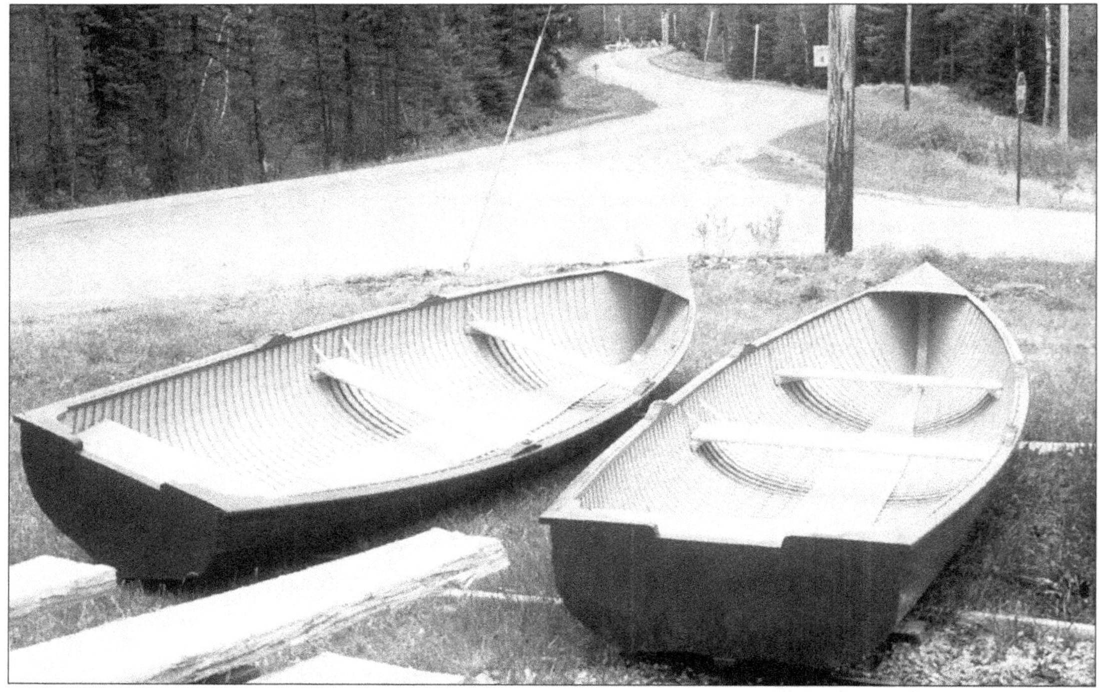

No. 2 and No. 3 transom Rangeleys outside Herb Ellis's shop
(Michael Chaney, Maine Folklife Survey)

when nine- and ten-horsepower engines were available. Not everyone thought that a ten-horsepower outboard belonged on even the widest, heaviest Rangeley, including Skeet Davenport, whose position was buttressed by experience:

> Most of your Rangeley boats don't take anything over a 9-horsepower, and a nine-horsepower I wouldn't recommend for steady use on a Rangeley boat. A six-horsepower and down is the best motor to run on.... The thing is, if the boat's new and sturdy, it's okay. But if you get a boat that's old and take a nine-horsepower. and snap it up full speed real quick, you can tear the transom out. In fact, I've seen two or three transoms tore out, the motor's down on the bottom, and you're sitting in a boat without—ready to swim. This is where the big motor is not good on a Rangeley boat.

Other Rangeley variations took form, too, also attributable to the outboard motor. Some boats were turned out at 20 feet or more, now that engines were pushing instead of arms pulling. Builder Martin Fuller made high-bowed Rangeleys expressly for use with a motor, and Harland Kidder saw them on the water.

> Yup, it was lapstrake and cedar, but the stern was built altogether different than any other Rangeley boat. It was low and it was undercut and, in fact, you had to watch it going downwind that you didn't get water over the back of it. But they were made strictly for an outboard—they swept up.

Another accommodation to the outboard was adding several more strakes to a boat, giving it a specific function. Mr. Davenport used such a Rangeley.

> I even had a Rangeley boat that I guided out of with a couple of extra strakes in it. It sat high in the water and was almost impossible to row, but it was a great boat to troll out of because you didn't get any splash in bad weather.

There was one means of altering the Rangeley boat that didn't depend on woodworking or mechanical systems. Paint could give a boat an identity, make it belong to a camp, club, or person. Herb Ellis recounted the varied paint schemes applied:

> We started out painting all boats all white inside and gray out. Then the old guides started complaining. You come off that lake at night and you can't even see. Well, you can't. It's like snow-blind. I said, "Why don't you have it gray?" Well, I painted some of them gray. That caught on so we shifted and they were all gray after that.... Now, like the Percy Club, they paint theirs light gray [inside] and dark gray out. I had one outfit that wanted theirs maroon outside. And we get special orders for different clubs. I built one that was gray inside, white outside, and back then, we used to get what we called varnish-stain. We

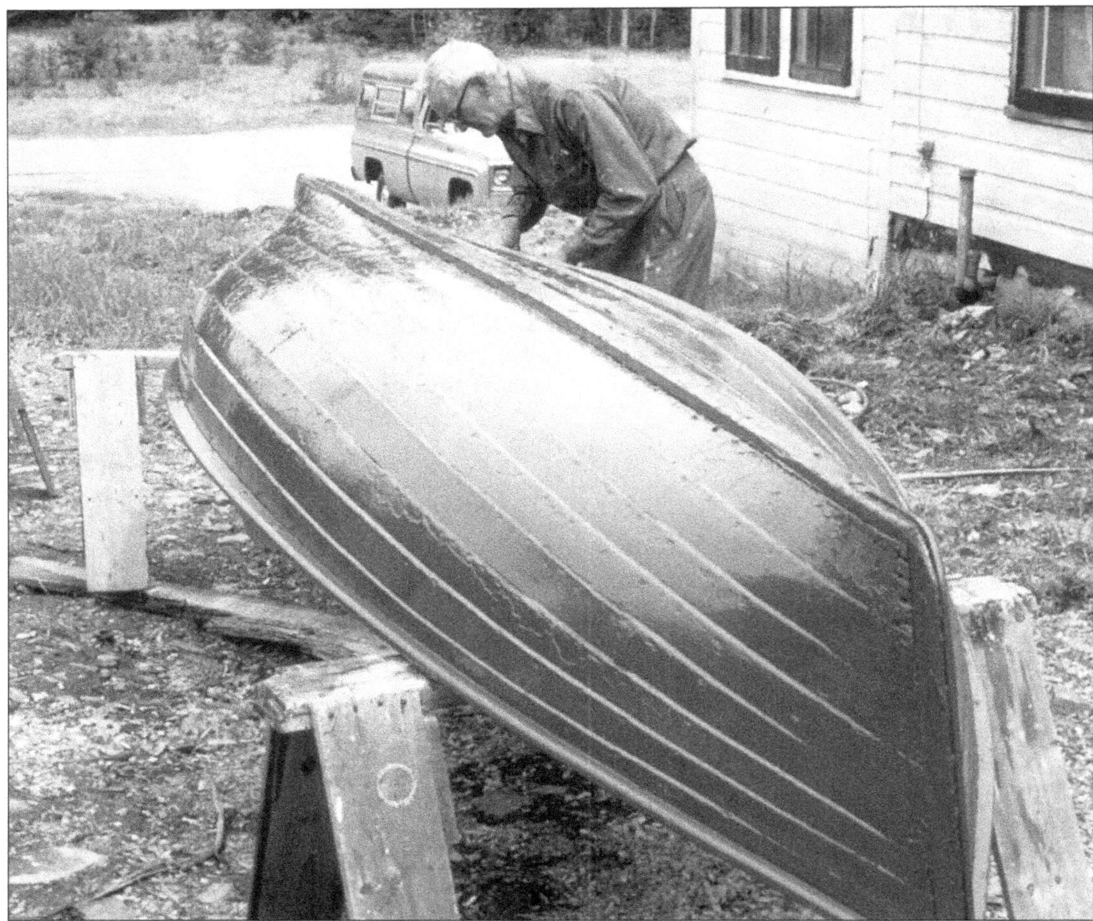

Herb Ellis applies the standard color—green *(Michael Chaney, Maine Folklife Survey)*

used the mahogany varnish-stain on the top of the deck. Made a beautiful-looking boat. Then the Oquossoc Angling Association, they used an Argentine orange inside and the outside was black, except for the sheerstrake, which was red. We paint them whatever they want, but the standard color was green.

Color schemes were not taken lightly. The Oquossoc Angling Association had decided in 1884 that all club boats would be painted black with a red stripe under the gunwhale, and requested that members have their boats painted similarly. But the value of Rangeleys in different hues went beyond clannish appeal. In cases of emergency, it allowed a boat and its occupants to be identified among the green Rangeleys on the water.

It would be wrong to think that the Rangeley boat never left its home, the lakes and ponds of western Maine. As it was developed early, its dispersion began early, and this craft with so many good qualities became part of sporting culture in new places. In the 1880s, disenchanted members of the Oquossoc Angling Association formed the Percy Summer Club on the shores of Christine Lake in New Hampshire. With them came a fleet of Rangeleys. Not far to the north, Rangeleys served anglers on Lake Megantic and Spider Lake, Quebec. The boats worked on Maine's Sebago Lake and the Belgrade lakes. Builders in Rangeley regularly shipped the boats to customers throughout New England, the Mid-Atlantic, and beyond. Mesmerized by the Rangeley and its region, broadcaster Lowell Thomas toted one around the Berkshires looking for bass. A customer had Herb Ellis make one up with copper fastenings, solely for striper fishing in the Cape Cod Canal. Ever loyal to the Rangeley, sports could not do without this stable but shapely boat for adventures close to home.

As Skeet Davenport plainly put it, "Originally you had your old aristocrat and the only thing there was to fish out of was the Rangeley boat." There is not much to be said about the time when—suddenly—there were things to fish out of besides a Rangeley. The postwar era brought forth boats that had never been a living thing and required little care. They were not made along the waterfront or in back-street shops in ones and twos; they were manufactured in plants by the thousands. They were not sold by reputation and word of mouth, but were advertised nationally. The usual craft seen on the Rangeley Lakes was, more and more, aluminum, and then fiberglass. The wooden Rangeley became the boat of traditionalists, of the past.

Only two shops remained to build Rangeleys after World War II, those of S. A. Collins & Son and Herb Ellis. Both sputtered along for years, often doing every kind of woodworking but boatbuilding, always doing more Rangeley repair than Rangeley construction. Herb Ellis, the last in the line of Rangeley builders, set his final boats outside the shop in 1981. A stroke made him turn to wood projects more manageable than boats of 17 feet. In the succeeding years, Herb Ellis's simple home and shop, seated on a rise in the road between Rangeley and Oquossoc Village, became a heritage destination. Museum curators, photographers, wooden boat builders, and revivalists showed up on the Ellis doorstep to record his words, tools, and work, all of which Herb delivered with a flinty, impatient intelligence. What got described was no-nonsense production boatbuilding, the standard that held from the Barretts and their contemporaries right down to Herb Ellis.

> There is a young man down around there, Camden, who is building Rangeley boats. He asked, "How long does it take you to build a boat?" Well, I said, "Well, if I have to put in more than ten or twelve days I lose money." He said, "God, I took over a year to build mine." I said, "Well, it makes a difference if you have been doing it one year or forty." No, I used to be able to start out with lumber and in twelve days I'd have your boat ready to row. But as I say, that was back when I worked on a dead run.

The "dead run" resulted in fifteen or twenty Rangeleys per annum, which Mr. Ellis allowed was "fair for one man."

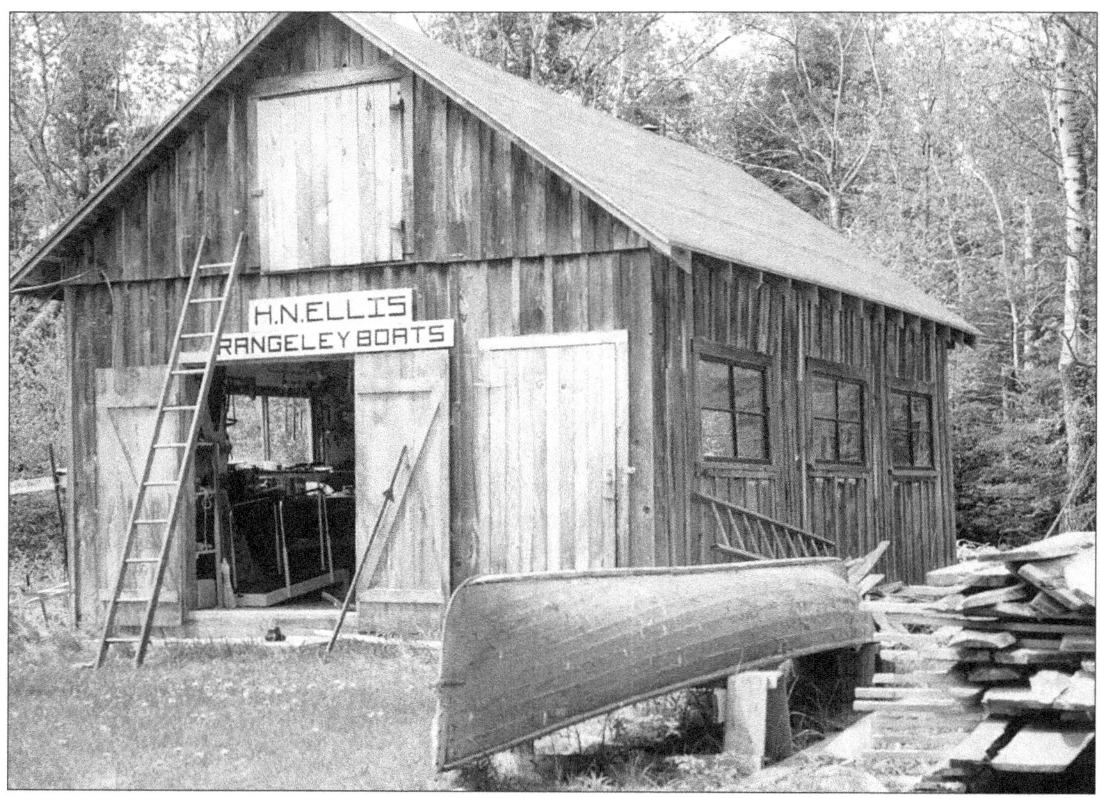

The Ellis Boatshop between Rangeley and Oquossoc *(Apprenticeshop of Atlantic Challenge)*

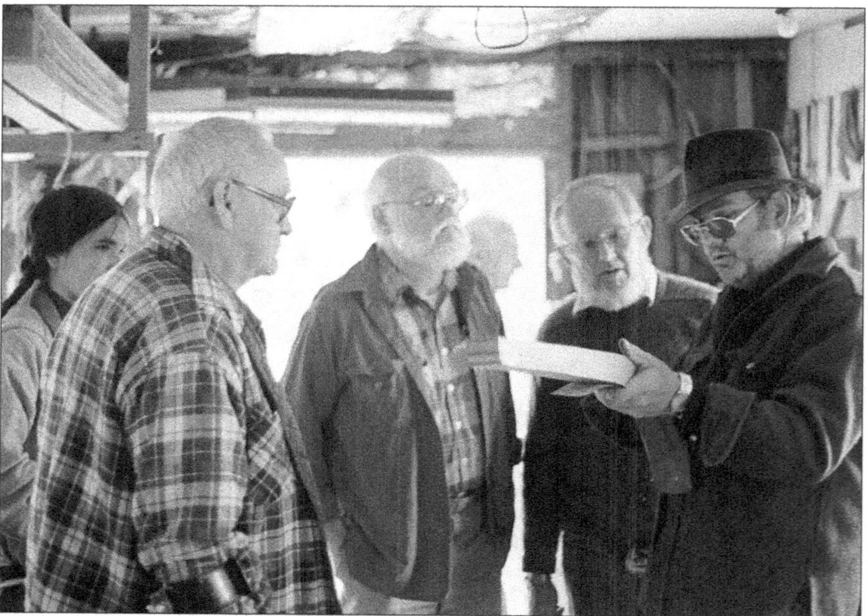

Herbert Ellis, at left in plaid shirt, with disciples in his shop
(Apprenticeshop of Atlantic Challenge)

Of course, back when they built double-enders and like that, Barrett would have a crew in the wintertime and they would put out twenty-five or thirty. I put out twenty-six one year alone. But there was a bunch of them double-ended. There is less work in building a double-ended. Not quite so many strakes and they are shallower and not as many ribs.

A near-frantic pace of building seemed to prevail in the Frank Barrett shop where Herb cut his teeth before becoming the successor.

This Frank, he was diabetic, kind of a nervous type and he did everything on a dead run, and of course when you work with a man like that, that is how I got in the habit of chain-smoking. You may think it is foolish with all the shavings around but we'd have three or four cigarettes the length of the boat. No matter where you were in the boat you could reach out and take a drag off a cigarette. Of course those were Depression days so we were smoking Wings. Ten cents a package. As I said, that is where I got the chain-smoking habit. After that I went rolling my own. When you roll your own and you lay it up it goes out, it don't keep on burning. As Frank said, this is the safest place in the world. If you drop a cigarette here, within three minutes those shavings would be ablaze, which you could see it and put it right out. He said otherwise something else would smolder for two or three hours and then catch on fire and you would be home when it would catch on fire. I guess he was right; I never caught the place on fire.

Herb Ellis's trial by near-fire took.

I worked for Barrett the summer of 1936, a little bit in 1937, and bought him out in the fall of 1938 and went ahead and built a boat. Made a few mistakes but I figured I would. I built it up, then had him come down and he showed me what was right and what was wrong. Then he said anytime all I had to do was holler. So, anytime I got stuck on anything I didn't know about, I would just go see him and he would straighten me out. That was it, but I learned by observing the summer and a half that I worked for him.

Herb's decision to purchase the Barrett Shop—patterns, tools, and a story-and-a-half building in town with a little lake frontage—for a borrowed $2,500 made economic sense, even in a remote Maine town at the end of the Depression. Rangeley's hotels and sporting camps were still intact, and the need for boat repair plus new boats kept the shop humming nearly year-round.

Herb built his boats on the hundred-year-old strongback and molds that came from the Barrett shop. He explained that the actual construction of a Rangeley boat could be reduced to these steps: Mark and saw the planking, bevel the planking, set the keel upside down on the strongback with the transom and stempost hanging from it, plank the boat over the molds, fastening the strakes with clenched nails, flip it right-side up and paint the interior, add the decking, attach the rails, then the ribs, and finally the cutwater. After that, it was the small stuff—a second interior coat of paint, thwarts, seats, oarlocks, exterior paint. In actual fact, the Rangeley was a fairly complex wooden boat to construct.

In 1986 a boat enthusiast named George "Bud" Brackett convinced a somewhat confined and ill Herb Ellis to guide him in making a Rangeley in much the same way that Frank Barrett had guided Herb in the late 1930s. The project took place over three September weeks in the boatshop now adjoining the Ellis home and produced, in addition to a Number One Rangeley with wineglass stern, fifteen single-spaced pages of notes on technique and construction. Bud Brackett wrote of Herb at the time, "His mind is right on the money. He hasn't forgotten one thing."

In what detail had Herb Ellis remembered the building of the Rangeley, the boat he'd been around for more than forty years and sometimes turned out at a pace of two dozen a season? The steps and their logic came so easily, it seemed as if he was still constructing them in his mind, even if his body was no longer able to.

The trickiest aspect of it was getting set up right. If you don't get the right curve to that backbone when you are setting up, then the planks don't go in right. You have to spring them too much. Some of them you have to spring anyway. That's what shapes them, makes them come

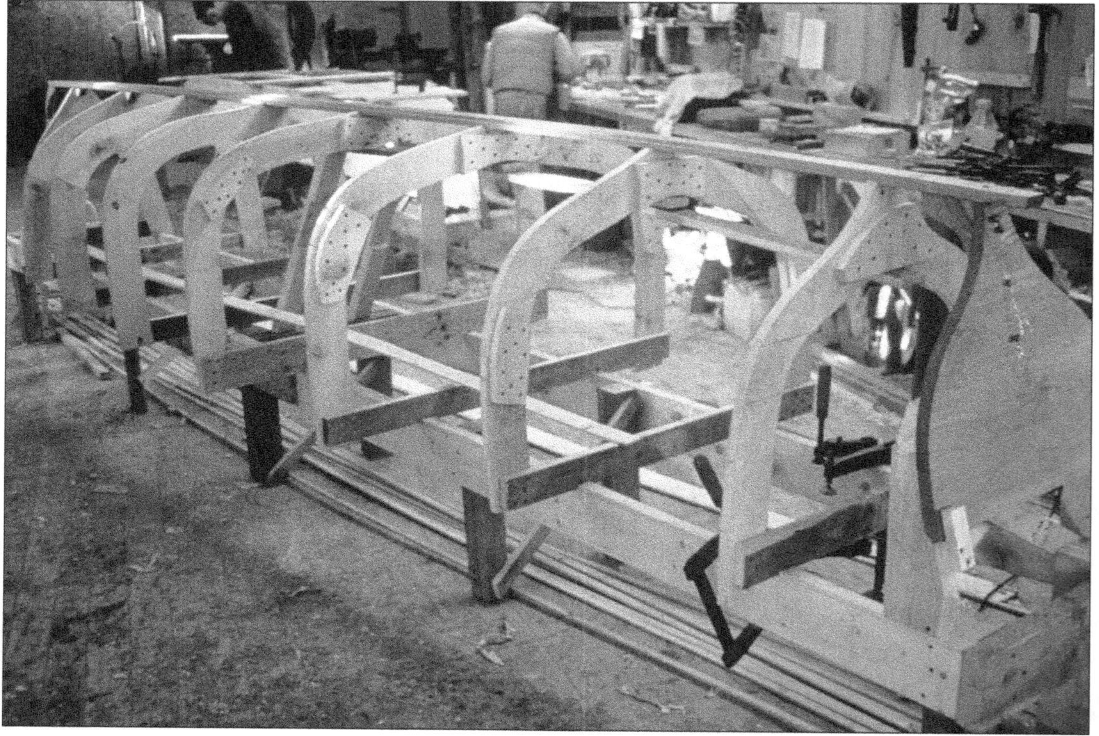

"Getting set up right." A Rangeley underway at the former Rockport Apprenticeshop. (*Apprenticeshop of Atlantic Challenge*)

around into the shape it's supposed to. But you don't want them to spring too much or you'll split something. I guess that's the most trickiest part of the boat. Your setup and your planking. Once you get that done, it's all clear sailing.... When we are building a boat, we start with a pile of lumber of an inch and an eighth thick. You lug a bunch of that into the shop. You have your patterns all laid out, shortest one going right up to the longest. You throw one of those boards onto the bench, look at it and see just about where you can get some good lumber out of it, and once you have got used to it you can look at your plank and you can know what plank you can get out of it. Put your pattern up and check, that's why you have them going from shortest to longest; that piece happens to be a bit longer you take the next longest one and if you haven't got quite enough you take the next shortest one. Lay it on and mark it and you make a mark on the pattern so's you can keep track of how many you got out, and we used to get out planking for fifteen to twenty boats at a time. It was faster and cheaper to do it that way.

On ribbing the boat, Herb Ellis's description was dramatic, almost visceral, and laced with anticipation of the work:

When I was younger, when I was in the business all the time, I put in about fifteen ribs an hour. Twenty-two nails to a rib, and you'd drill a hole for every nail and you'd stick every nail and then drive it off. You should be able to rib a boat in about four hours. There are sixty-seven ribs. Well, you're not standing there! When I was a lot younger I worked on a dead run, and didn't stop for anything. At the last of it I was taking about six hours to rib a boat, whereas we used to do it in about four. I stopped running and walked. You get up to fifty and sixty, you don't run so much anymore.

You go out in the morning and start your ribs heating. I used to go when I got up, light the gas plate, then by the time I got my breakfast and got out there it would be pretty warm because I would put them in and let them soak all night. They'd be pretty warm but they wouldn't be warm enough to go into the ends where they would sharp bend. So, you start out in the middle of your boat and you're in there for about an hour or so. By that time your ribs are really hot. You could just about touch them. Once you get your hands toughened into it you can, but at the first of it, it's kind of hot. I never wore any gloves, you have to have the feel. You can't handle nails with anything on. Then the bow is very sharp, you put them back to back about seven or eight ribs. You take on the back side of them and shave them away. Then you take them out, you just take them out between your thumbs and fingers on both hands; you pre-shape them with your fingers. Then you slide them in and you tap them from both sides until it is tight. Then it is a one-piece rib, throughout the hull. You can break a lot of them if you don't have them hot enough. It's almost a square bend. But you get those hot enough, 'cause we pick the straightest grains we can find to do it, you shave them away on the back and you can pre-bend them. Once you get them all done, you take the belt sander and you put your corner blocks in, put your inwales in, and take your sander and you go right across, crossways between the inwale and the rib end; the planking at the bow, you cut that smooth-straight, then you take a fine belt and you go lengthwise on it, smooth it, right up to a polish.

As the boat was nearly perfect from the start in form and function, it defied most innovation, by Herb or other builders.

I never made any other changes to the boats except the flat seats instead of the round. Not much to change on them. When you got something good leave them alone, they always told me. We had patterns for the stern, your stems, cutwater. You had patterns for your knees, corner blocks. Course if you changed the stern you had to change your corner block. You had patterns for your deck and transoms and of course your stems—your bow stem and your stern stem. You have a pattern for your keel piece. I'd say there was a pattern for every piece in there.

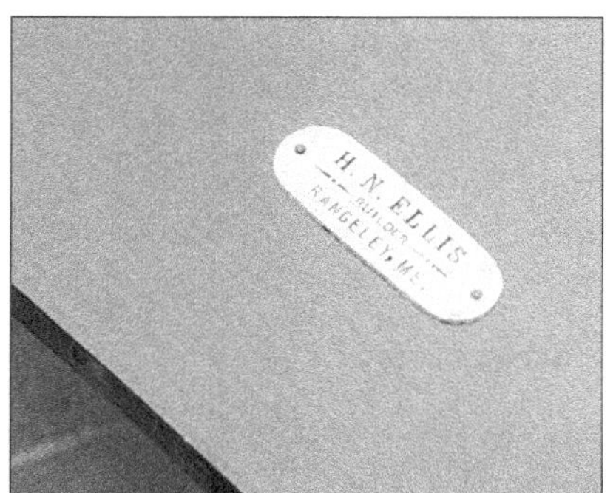

The Ellis nameplate at a Rangeley's bow *(Michael Chaney, Maine Folklife Survey)*

Herb Ellis died in 1997, but the Rangeley boat has survived. It lives on in places where tradition is respected, and some nostalgia admitted to. "The managers of the Oquossoc Angling Association have retained all the semi-aboriginal character in their camps, grounds,

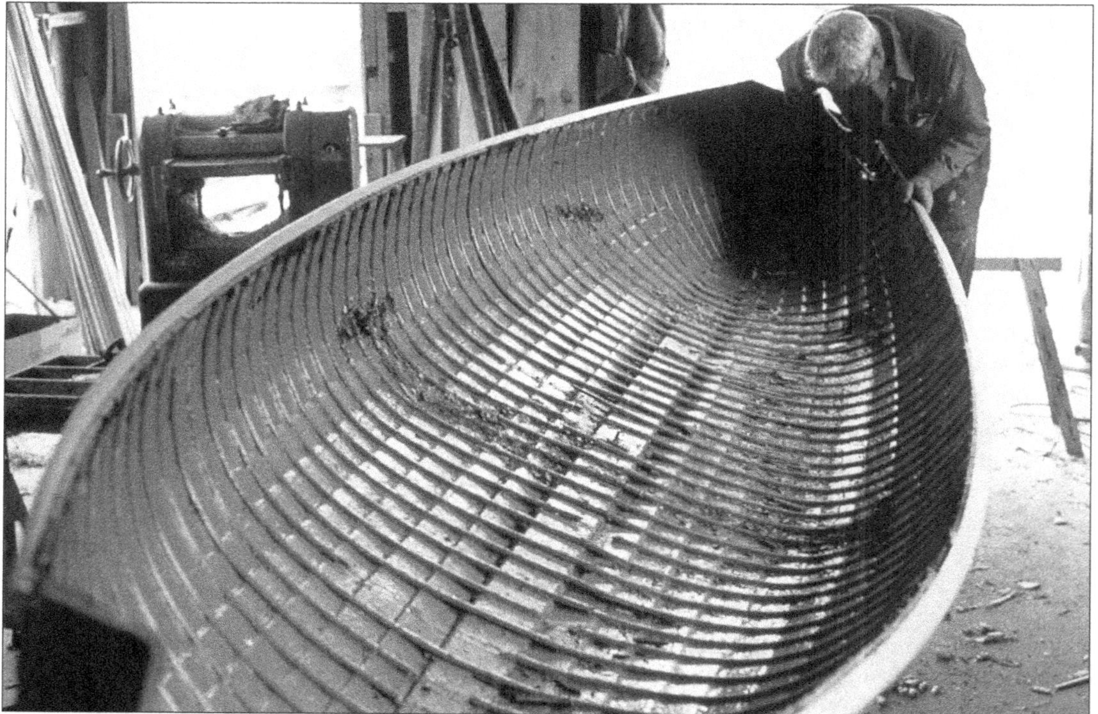

The Rangeley's innumerable ribs—Herb Ellis repairs a boat
(Michael Chaney, Maine Folklife Survey)

and appointments." So wrote a guidebook author in 1887, and it remains true today. This private enclave says to the visitor that it is not about socializing or golf, it is about catching fish. For more than a century, politicians, medical men, captains of industry, and sportswomen have arrived not for publicity or notoriety, but for trout and salmon. The O.A.A., fountainhead of Rangeley's fishing culture, is a stronghold of Rangeley boats.

In a car barn made over for boats lie the black-hulled Rangeleys of the Oquossoc Angling Association's members. There are thirty or so,

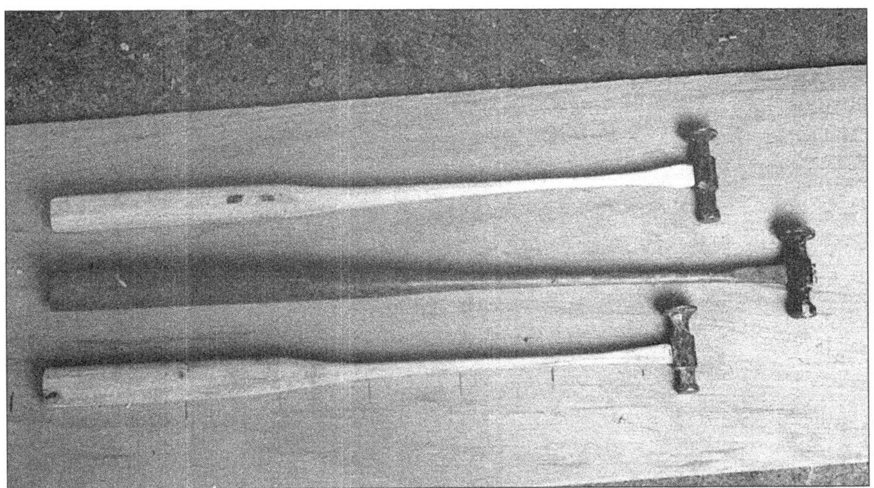

Tools from the Ellis Shop: hammers for clenching nails, with hand-hewn, hornbeam handles (*Apprenticeshop of Atlantic Challenge*)

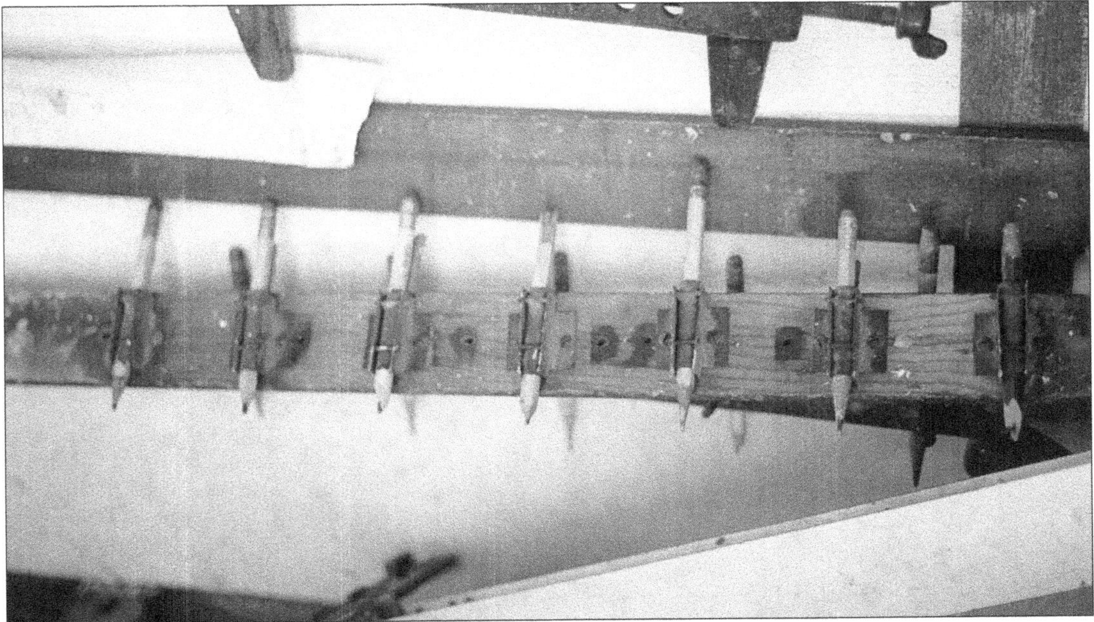

Tools from the Ellis Shop: a seven-pencil gauge for marking nail spacing on planks, a Barrett Shop invention (*Apprenticeshop of Atlantic Challenge*)

double-enders, ones, twos, and threes. The work of a number of builders is present—a Charles and Thomas Barrett boat, several Arnburgs, numerous craft by Frank Barrett, Herb Ellis, and others. This may be the region's best and largest collection of indigenous boats. The Rangeley thrives here because members enjoy fishing from it as they always have done, and because each one is routinely repaired and painted and sheltered for winter. It is likely there will always be Rangeleys at the Oquossoc Angling Association docks.

Much the same enthusiasm toward Rangeley boats prevails at Grant's Camps, on the eastern shore of Kennebago Lake. A haven for anglers since 1904, Grant's is where the late Walter "Skeet" Davenport concluded his career in the outdoors. For twenty years, Mr. Davenport was part-owner and resident manager. He put the attitude of Grant's Camps fishermen toward boats quite simply:

> You look at Grant's Camps, and they've got one or two metal boats there, but they're only there for last resort. All the rest of the boats are Rangeleys, and they rent those first. If you had the fifteen Rangeleys sitting there and a guy came in to hire a boat, he isn't going to rent a metal boat. You might say, "Well listen, why don't you take this one?" He won't take it, he wants the Rangeley boat. Because that goes along with the mystique of the place.

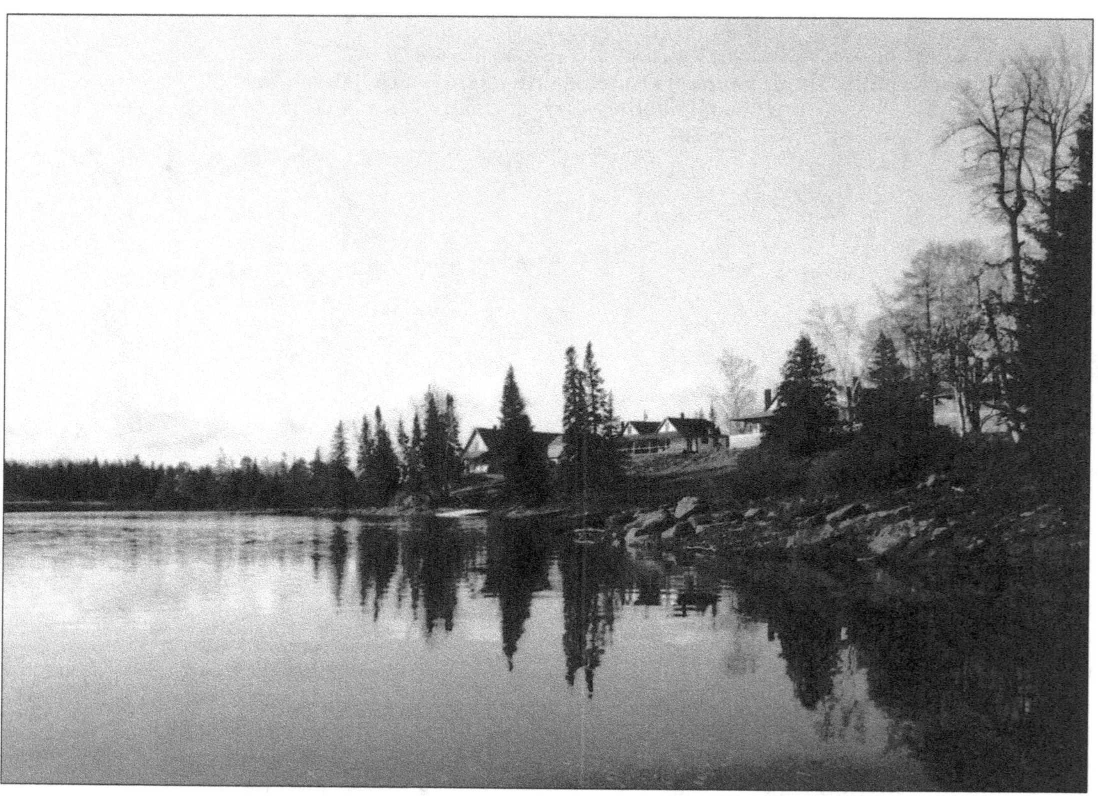

　　The Oquossoc Angling Association from the Kennebago River *(Stephen A. Cole)*

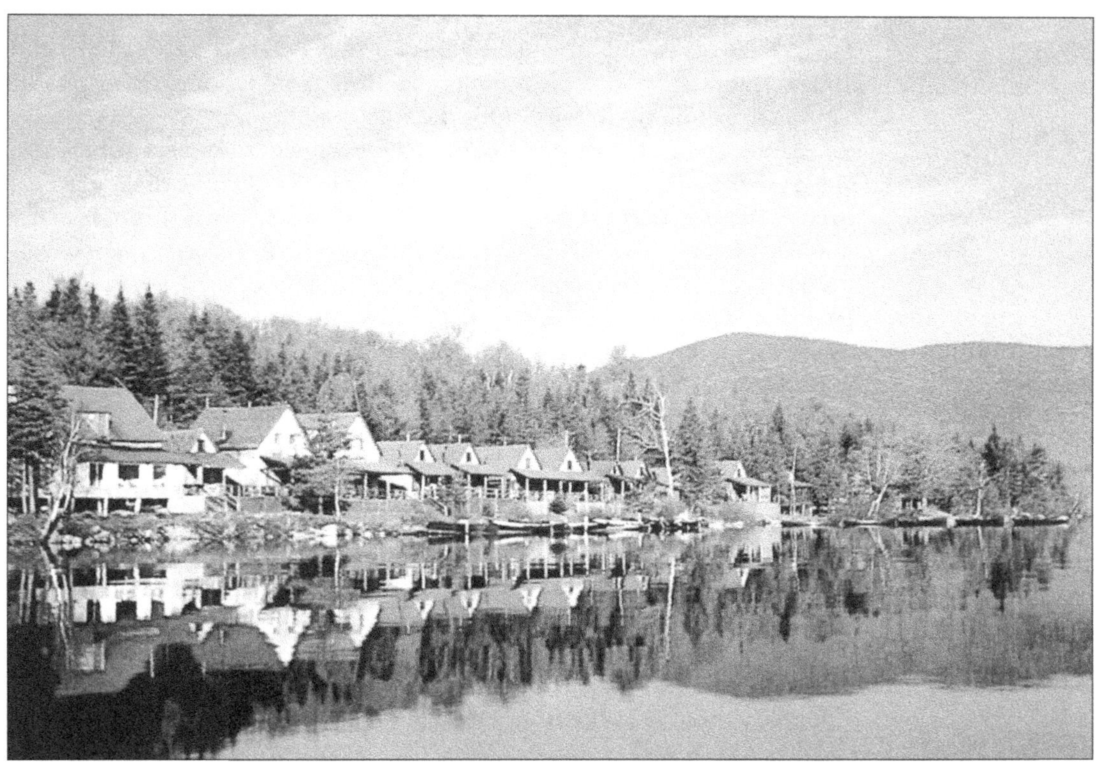

Grant's Camps, Kennebago Lake *(Margaret S. Davenport)*

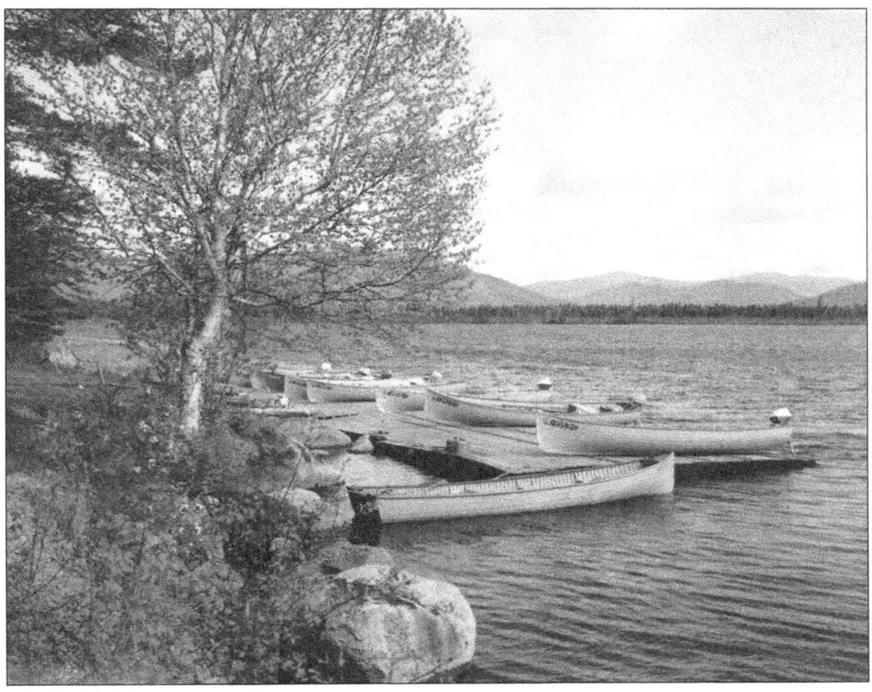

The Rangeleys at Grant's Camps *(Margaret S. Davenport)*

It is a mystique that, besides wooden boats, includes moose feeding in marshy spots called "logans," pools known as Drinkwater's Bath and Cousin John's, and the pleasant realization that, as one sport put it, "There is nothing to do but fish or die."

Along the waterfront at Grant's, the Rangeleys wait for fishermen, fended off from the docks with old tires. Some boats still have the traditional round seats, others sport fly seats. The floorboards, worn through many times, are now plywood sheets. Where the guide sat in years past is bow ballast consisting of a plastic milk box filled with cobbles. These are Rangeleys that work. They have been painted so many times that the bow-plates are now covered, the names of the builders obscured. Because customers favor the Rangeley, Grant's Camps maintains its fleet the best it can and buys up used Rangeleys when they can be found.

The fishing clubs and sporting camps of western Maine are the last best refuge of the Rangeley boat. There, the boats are prized as much for the aura they evoke as for their competence on the water. Traveling the lakes, one will see the occasional handsomely painted Rangeley on the dock of a private cottage. But one will see far more moldering in the underbrush along the shore or slowly coming apart under porches and in boathouses.

Nevertheless, Skeet Davenport believed a small boatbuilder could still make a go of it in the Rangeley's birthplace.

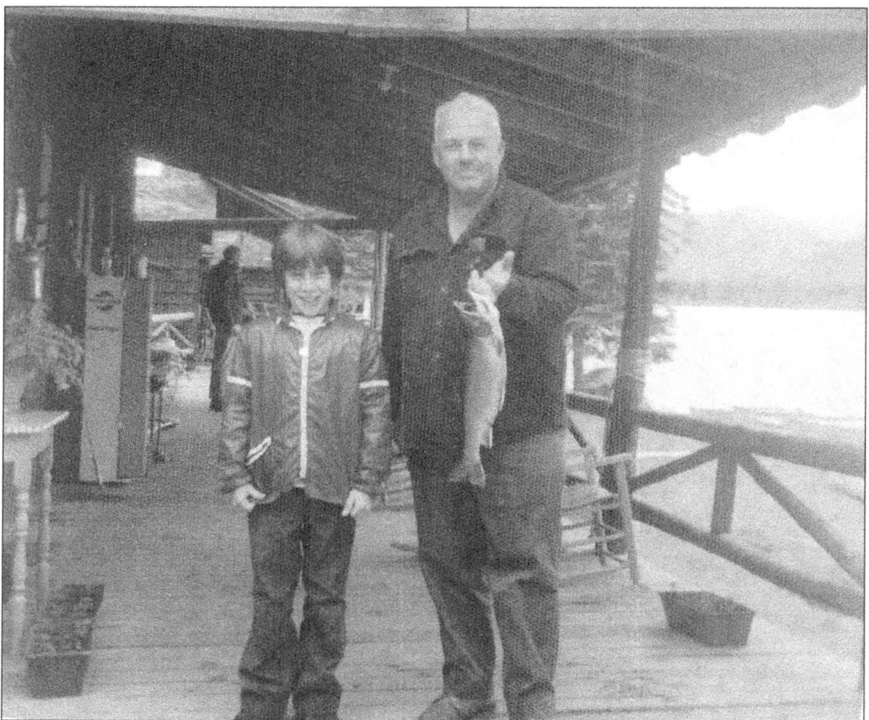

Young Mike Koob of Oquossoc and Walter Davenport at Grant's Camps
(Margaret S. Davenport)

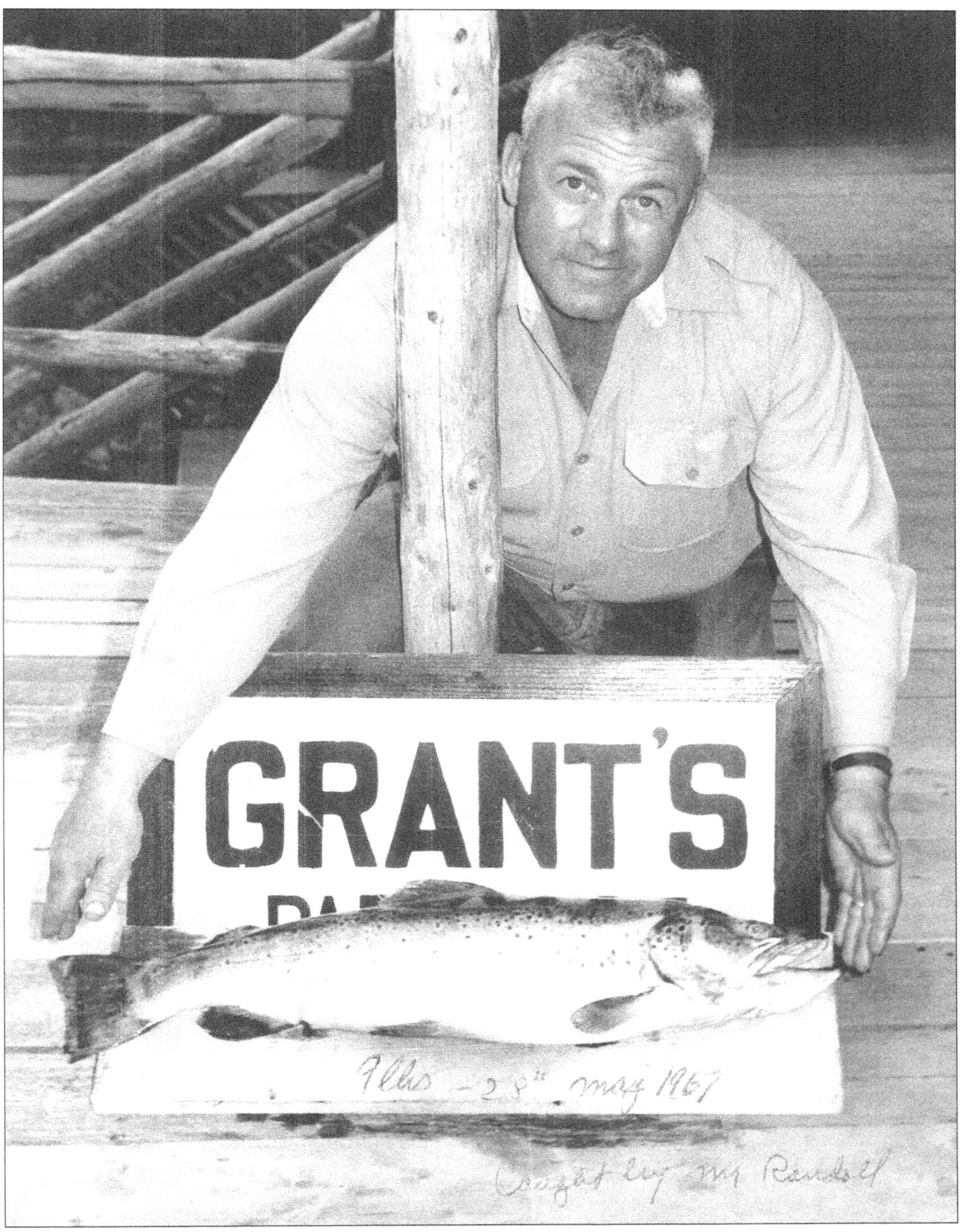

Walter "Skeet" Davenport and a sport's salmon, 1967 *(Margaret S. Davenport)*

He's got to be competitive, he's got to have a good boat. Not only does he have to make the boat, he's got to be in a position where he can repair it. He's got to be able to service it. I think that every camp here would buy a Rangeley boat providing they don't have the price so high for that type of boat.... I think where the Rangeley boat can be sold is to the fly-fisherman. Fly-fishing has had a big resurgence, there's a lot of fly fishermen now. You can tie the two things together. If you promote, you almost say this is the fly-fisherman's boat.

Only time will tell whether the Rangeley boat enjoys a renaissance in Rangeley, Maine. For the past few years boatbuilder Richard Woodward has turned out several Rangeleys in the region and taken on some Rangeley repair work. His boats are the 17-foot, Number One transom model, constructed in marine plywood and based on the lines that traditional boat guru John Gardner took off a 1930s-era Herb Ellis craft. More than a hundred years ago, the *Rangeley Lakes* newspaper promised and awarded a Rangeley boat to the guide who sold the most subscriptions of their fledgling journal. In an echo of these times, Maine's Department of Inland Fisheries and Wildlife is now raffling off one of Woodward's classic green Rangeleys to the purchasers of 2007 hunting and fishing licenses, deer permits, and moose hunt lottery chances.

Whether this current resurgence holds or not, it seems there will always be Rangeley loyalists, men like Skeet Davenport, who never fished more than a half-dozen days on the lakes in anything but a Rangeley boat. He loved not only its form, but its function.

I like to be right down next to the water when I'm fishing. When I'm in a big boat, I don't have the faith in what I'm going to catch out of a big boat. I know that it's probably all in my head.... This boat, you can quietly go into the shallow water—you know, you don't disturb a lot of fish. The big boats have revolutionized the actions of your fish. You get twenty-five or thirty big boats milling around a school of fish and they're right on the shore—where they used to stay there in the old days—they'll drive them right back out into deep water. I like it because if I find fish in shallow water, I know that if a lot of big boats don't come in and drive them out, the fish are going to stay there. It's just something you got faith in and I guess if you're stubborn, why, you don't change. 'Course, it's a lot more comfortable to fish out of the big ones. Set there, have a nice soft seat and a windshield, a cover over the top of you, this sort of thing. But when I get to the point—personally—when I have to have that, well, then I'll stay ashore.

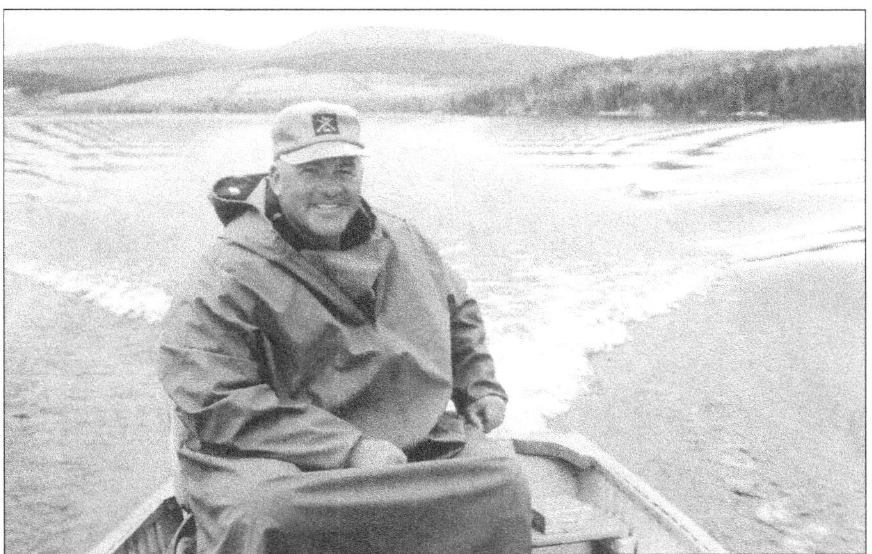

Walter C. "Skeet" Davenport *(Margaret S. Davenport)*

(Apprenticeshop of Atlantic Challenge)

Postscript

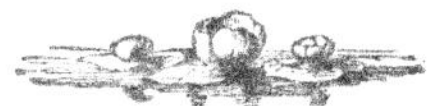

It comes as a shock as one drives into western Maine from the coast. There is still ice on the lakes in May. Here in Rangeley—nearly halfway between the equator and the North Pole—spring has arrived with little certainty. Precious few flowers are showing and the buds on the trees are small, tight, and red. The only birds to be heard are robins, heralding the promise of the season once more. Shelves in the Main Street shops are meagerly stocked; perhaps only half of the fishing tackle that will eventually grace the hardware store's pegboard is in. The town and its people, like the trees and birds, are just reawakening to life out-of-doors. The ice has shrunk away from the edge of Rangeley Lake, but the fish haven't really moved to the shore yet. Perhaps with a west wind, the ice will break up.

Walking down Richardson Street in earlier years, you would have seen Walter "Skeet" Davenport, fisherman and former guide, out in the yard with his Rangeley boat up on sawhorses, renewing the paint on its gray hull and red gunwales. He would have lubricated the outboard, already, and even cast a few streamer flies in Rangeley Lake's thin margin of water. Skeet Davenport has died, but other eager spring anglers have taken his place.

In a worn and warm house elsewhere in town, Bud Wilcox would have been working under strong light in a back room, tying streamer flies for the coming season. He'd have brought out lovely, lurid flies on wooden trays: size fours, sixes, tens, and twelves. They would have contrasted wonderfully with his workingman's pants and shirt of dark green twill. Mr. Wilcox's hands were big and rough and shook just a little, yet his flies were small and perfect. A number of the streamers were his own design, including the marvelous Kennebago Smelt and jaunty Tri-Color. Perhaps he'd have fashioned a new fly for the new year, like the Long Pond Minnow, red for its tummy full of roe, buff-green-brown for its sides. Knowing where you'll fish, he would have suggested some patterns and assured you that the logans ought to hold some trout. As for himself, he'd wait until the ice was truly out of these waters before venturing after landlocked salmon and brook trout. Bud Wilcox is gone, too, but fly tiers like Gray Wolf now continue the legacy of Mrs. Dill, Carrie Stevens, and Mr. Wilcox.

—•◆•—

At the town landing, there are about twenty acres of ice-free lake. One loon is on it and one daffy Labrador retriever is in it. On this cool day, they will be joined by a watercraft with but a short history on the Rangeley Lakes. Even now, the windsurfer, clad in black synthetic armor, is preparing his sail and board. The board name seems to sum up the scene: *Fanatic.* Slowly departing the shore, the adventurer is watched incredulously by an elderly gentleman in an idling car. The sail is a geometric collage of hot pink, lime green, yellow, aquamarine, and purple, and on this early May day, it is easily the most colorful object in town.

In the village called Oquossoc, just a few miles beyond Rangeley, spring has its own signs. Every evening now, deer are coming out of the woods to feed in the fields just greening up, and the invisible peepers are singing for love. By day, the marina workers are moving Boston Whalers and other fiberglass models out of storage and washing them. The owners will arrive soon.

At the outlet of Rangeley Lake, near the village, spring seems less tentative and a person catches up with the dramatic phenomenon known in Maine as ice-out. This ice is no longer static, but active, and surges toward the cement pilings and steel grate of the outlet. The structure suffers slight but constant tremors. It is plate tectonics in miniature, as chunks of ice six feet across shift and shear, breaking apart and coming together in new ways. Flows jostle for position to exit the lake. Some are broken to precisely the dimensions of the outlet itself, then go rushing downstream to the falls. The sound is of crashing and shattering, like dozens of lead glass goblets dropped one after another. The stretch of water above the outlet may be clear one minute and entirely ice-bound the next as a new flotilla approaches death and dismemberment. There can be no doubt: The ice is in decline. It is pooled with water and shot with cracks and fissures. The final legacy of winter, the last holdout, is coming apart.

Ice-out has special meaning for the fishermen of the Rangeley Lakes, and it is for them both an event and a time. The best fishing of the entire year comes with the spawning of smelt, which occurs with a full moon in late April or early May. As the smelt head for spawning beds in the streams that feed these lakes, trout and salmon crowd the inlets and take them. When ice-out coincides with the spawning run, a fisherman casting streamers to imitate smelt will find incredible sport and big fish. But ice-out fishing is elusive. In 1989 the smelt spawned while the lakes were still locked in ice, and no one got the game fish. In twenty years of trying, a devoted Rangeley angler from away has only once arrived in town to find open water and spawning smelt. Most fishermen coming to Rangeley never expect to hit the smelt spawn and only ask over the phone, "Is the ice out?" "Is there open water?" "In places," the B&B owner answers. Quimby Pond is often the first open body of water, and a local guide has presented this bed-and-breakfast hostess with a lovely foot-long brook trout. Slit under the gills and down the middle, its head is cocked up and its entrails gone.

After the feeding frenzy of earliest spring, the salmon and trout quiet down and fishing is slow for awhile. On the shore of Rangeley Lake, adolescents smoking cigarettes are catching spawning perch and suckers and ripping the hooks from their mouths while grunting obscenities. At the town dock, other bait fishermen are pulling in small trout (which are rumored to be hatchery-raised) and reacting as if they are trophy fish. But for the gentlemen fly-casting streamers from a boat offshore, there are neither trout nor salmon.

At a lake more remote than this, a man in waders paces the shore in frustration. He has a new boat to launch and he wants to fish, but the ice is just breaking up and jagged flows weighing a ton or more are moving by his dock. A launching now would mean scraped paint, or worse, some planking stove in. He pictures being kicked in the ribs. An earnest fly-fisherman, his new boat is a Rangeley. For years he fished out of rented Rangeleys and imagined his own. Bringing the Rangeley from a coastal boatshop home to the lakes in mid-April, he found snow and sub-zero temperatures. Now it is early May and he wants his boat in the water. But the launching will have to wait until time and weather comply. Such are the vagaries of ice-out, and they are no different now than in 1889. Even then, anxious men from away bided their time while the ice departed, thinking only of their Rangeleys waterborne and of catching fish.

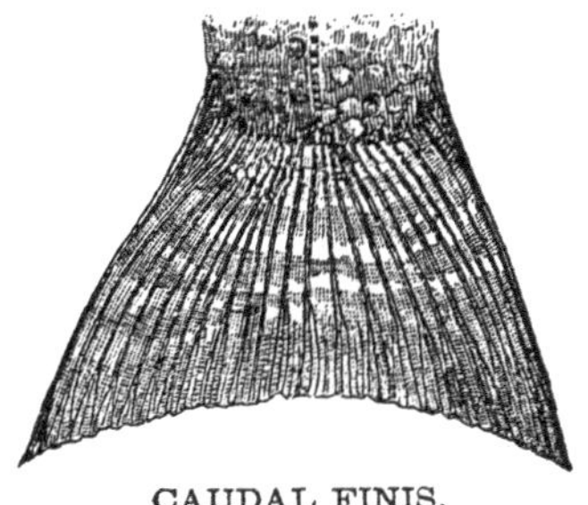

CAUDAL FINIS.

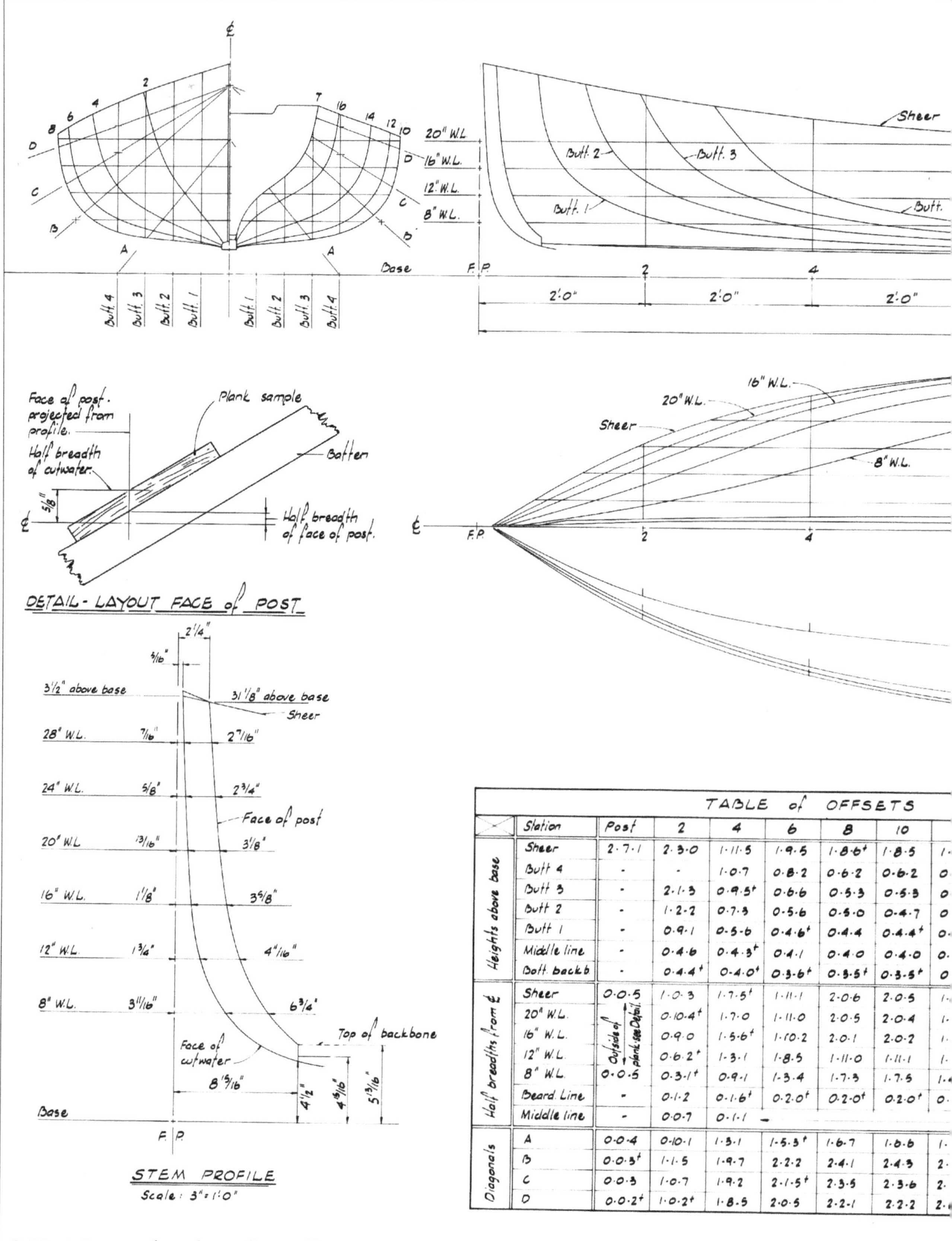

	Station	Post	2	4	6	8	10
Heights above base	Sheer	2·7·1	2·3·0	1·11·5	1·9·5	1·8·6†	1·8·5
	Butt 4	-	-	1·0·7	0·8·2	0·6·2	0·6·2
	Butt 3	-	2·1·3	0·9·5†	0·6·6	0·5·3	0·5·3
	Butt 2	-	1·2·2	0·7·3	0·5·6	0·5·0	0·4·7
	Butt 1	-	0·9·1	0·5·6	0·4·6†	0·4·4	0·4·4†
	Middle line	-	0·4·6	0·4·3†	0·4·1	0·4·0	0·4·0
	Bott. backb.	-	0·4·4†	0·4·0†	0·3·6†	0·3·5†	0·3·5†
Half breadths from ₵	Sheer	0·0·5	1·0·3	1·7·5†	1·11·1	2·0·6	2·0·5
	20" W.L.	Outside of plank see Detail.	0·10·4†	1·7·0	1·11·0	2·0·5	2·0·4
	16" W.L.		0·9·0	1·5·6†	1·10·2	2·0·1	2·0·2
	12" W.L.		0·6·2†	1·3·1	1·8·5	1·11·0	1·11·1
	8" W.L.	0·0·5	0·3·1†	0·9·1	1·3·4	1·7·3	1·7·5
	Beard Line	-	0·1·2	0·1·6†	0·2·0†	0·2·0†	0·2·0†
	Middle line	-	0·0·7	0·1·1	—		
Diagonals	A	0·0·4	0·10·1	1·3·1	1·5·3†	1·6·7	1·6·6
	B	0·0·3†	1·1·5	1·9·7	2·2·2	2·4·1	2·4·3
	C	0·0·3	1·0·7	1·9·2	2·1·5†	2·3·5	2·3·6
	D	0·0·2†	1·0·2†	1·8·5	2·0·5	2·2·1	2·2·2

Rangeley Lake Boat by Herbert N. Ellis, "Herb Ellis No. 2"

17' 2" x 4' 2"   Drawn by David W. Dillion

Plan No. 1: Lines, offsets, construction details   1.5" = 1' (original)   4/1987

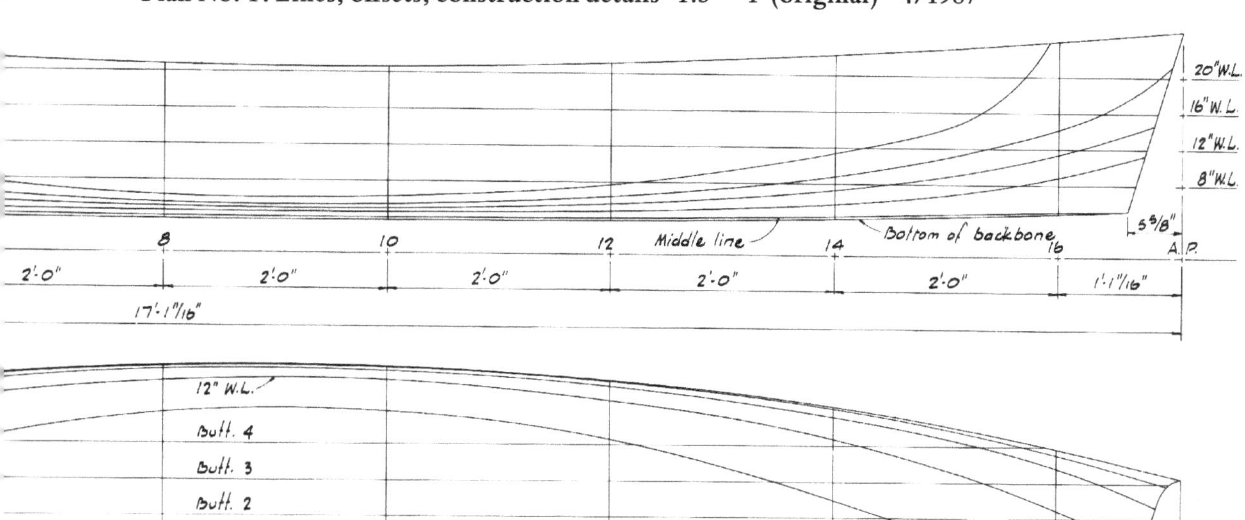

	16	Transom
2	1·11·7	2·1·1
3	-	-
6	1·2·1	1·9·3
2	0·11·2⁺	1·2·5
5⁺	0·8·6⁺	0·11·3
3	0·5·0	-
0⁺	0·4·6	0·5·1
3	1·3·6	1·0·4
1	1·3·1	0·11·5
2	1·1·5	0·9·3
0	0·9·0	0·4·5
7⁺	0·3·1	0·1·6
5⁺	0·1·3	0·1·1
—	0·1·1	-
	0·11·3⁺	0·9·1
	1·6·2	1·2·5
6	1·5·4	1·2·0
5	1·4·4	1·1·1⁺

Note that as shown in the profile that the middle line curves downwards as it nears the ends of the backbone. This is because the cut for the rabbet is made to give a full breadth of 2¼" for its whole length, and then tapered with no further cutting of a rabbet "notch" so that at the post and transom, the backbone is very nearly square-edged.

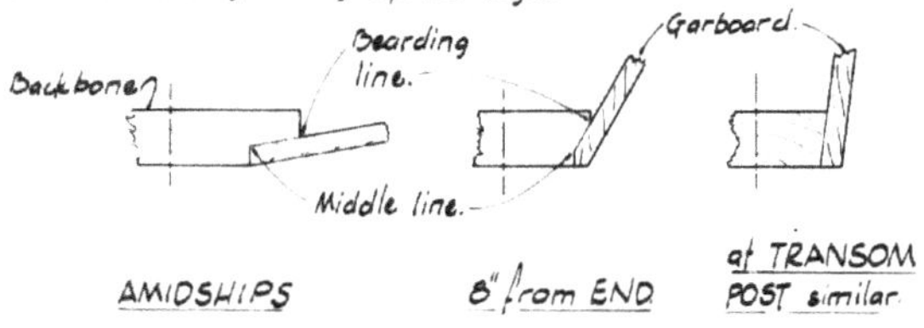

Offsets given in feet-inches-eighths to the inside of lapstrake plank.

Waterlines 8", 12", 16" and 20" above base.

Buttocks 4", 8", 12" and 16" out from ₵.

Diagonals: A up 20" on ₵, 16" out on base; B 28" up on ₵, 22" out on 8" W.L.; C 28" up on ₵, 18" up on Butt. 4; D 28" up on ₵, 22" up on Butt. 4.

Measured at Rockport, Maine in April 1987 by D.W. Dillion.

This boat was built at the shop of Herbert N Ellis in Rangeley, Maine during September, 1986, by Bud Brackett working under Herb's supervision and guidance.

The measuring of the boat and the production of these drawings was made possible by a donation from George E. "Bud" Brackett 1K.

The purchase of these plans entitles the purchaser to build one boat only.

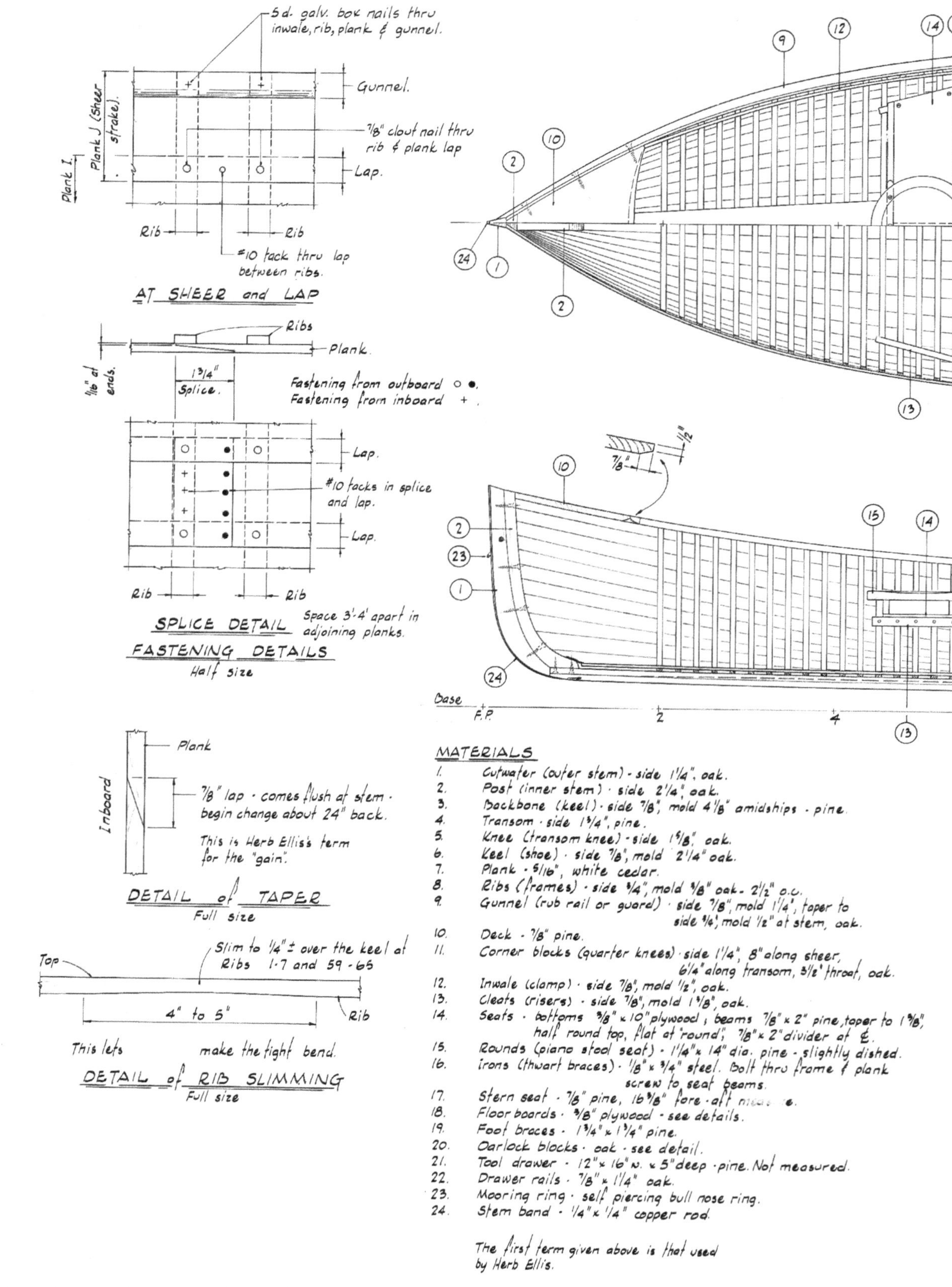

MATERIALS

1. Cutwater (outer stem) · side 1¼", oak.
2. Post (inner stem) · side 2¼", oak.
3. Backbone (keel) · side ⁷/₈", mold 4⅛" amidships · pine.
4. Transom · side 1³/₄", pine.
5. Knee (transom knee) · side 1⅝", oak.
6. Keel (shoe) · side ⁷/₈", mold 2¼" oak.
7. Plank · ⁵/₁₆", white cedar.
8. Ribs (frames) · side ¾", mold ⅜" oak · 2½" o.c.
9. Gunnel (rub rail or guard) · side ⁷/₈", mold 1¼", taper to
        side ¾", mold ½" at stem, oak.
10. Deck · ⁷/₈" pine.
11. Corner blocks (quarter knees) · side 1¼", 8" along sheer,
        6¼" along transom, 3½" throat, oak.
12. Inwale (clamp) · side ⁷/₈", mold ½", oak.
13. Cleats (risers) · side ⁷/₈", mold 1⅝", oak.
14. Seats · bottoms ⅜" × 10" plywood ; beams ⁷/₈" × 2" pine, taper to 1⅝",
        half round top, flat at "round"; ⁷/₈" × 2" divider at ₵.
15. Rounds (piano stool seat) · 1¼" × 14" dia. pine · slightly dished.
16. Irons (thwart braces) · ⅛" × ¾" steel. Bolt thru frame & plank
        screw to seat beams.
17. Stern seat · ⁷/₈" pine, 16⅝" fore-aft measure.
18. Floor boards · ⅜" plywood · see details.
19. Foot braces · 1³/₄" × 1¼" pine.
20. Oarlock blocks · oak · see detail.
21. Tool drawer · 12" × 16" w. × 5" deep · pine. Not measured.
22. Drawer rails · ⁷/₈" × 1¼" oak.
23. Mooring ring · self piercing bull nose ring.
24. Stem band · ¼" × ¼" copper rod.

The first term given above is that used
by Herb Ellis.

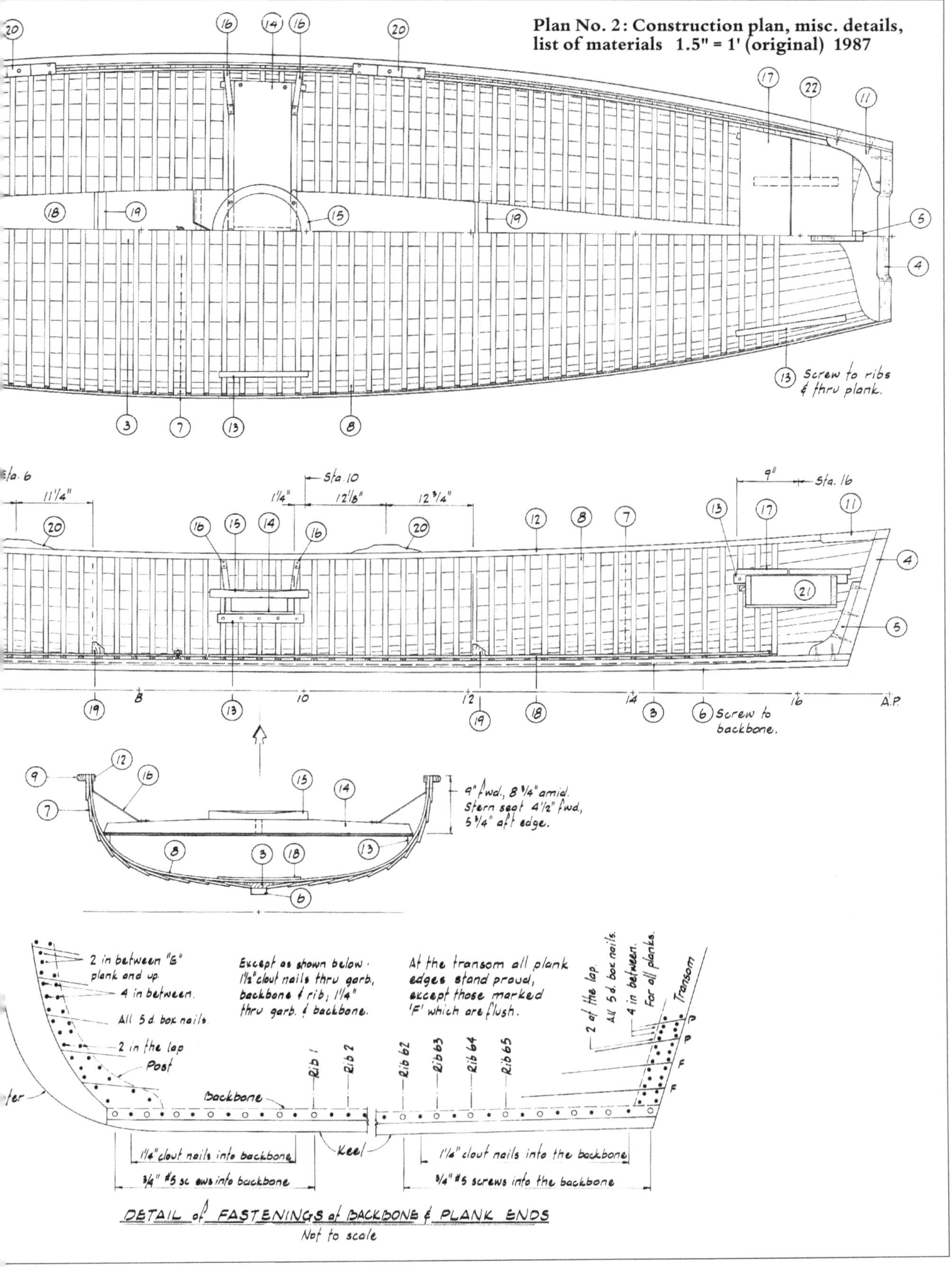
Screw to ribs & thru plank.
Sta. 6
Sta. 10
Sta. 16
11 1/4"
1 1/4"
12 1/8"
12 3/4"
9"
Screw to backbone.
9" fwd., 8 3/4" amid.
Stern seat 4 1/2" fwd,
5 3/4" aft edge.
A.P.
2 in between "B" plank and up.
4 in between.
All 5 d. box nails.
2 in the lap
Post
Except as shown below.
1 1/2" clout nails thru garb,
backbone & rib; 1 1/4"
thru garb. & backbone.
At the transom all plank
edges stand proud,
except those marked
'F' which are flush.
2 of the lap
All 5 d. box nails.
4 in between.
For all planks.
Transom
Rib 1
Rib 2
Rib 62
Rib 63
Rib 64
Rib 65
Backbone
Keel
1 1/4" clout nails into backbone
3/4" #5 screws into backbone
1 1/4" clout nails into the backbone
3/4" #5 screws into the backbone
DETAIL of FASTENINGS of BACKBONE & PLANK ENDS
Not to scale

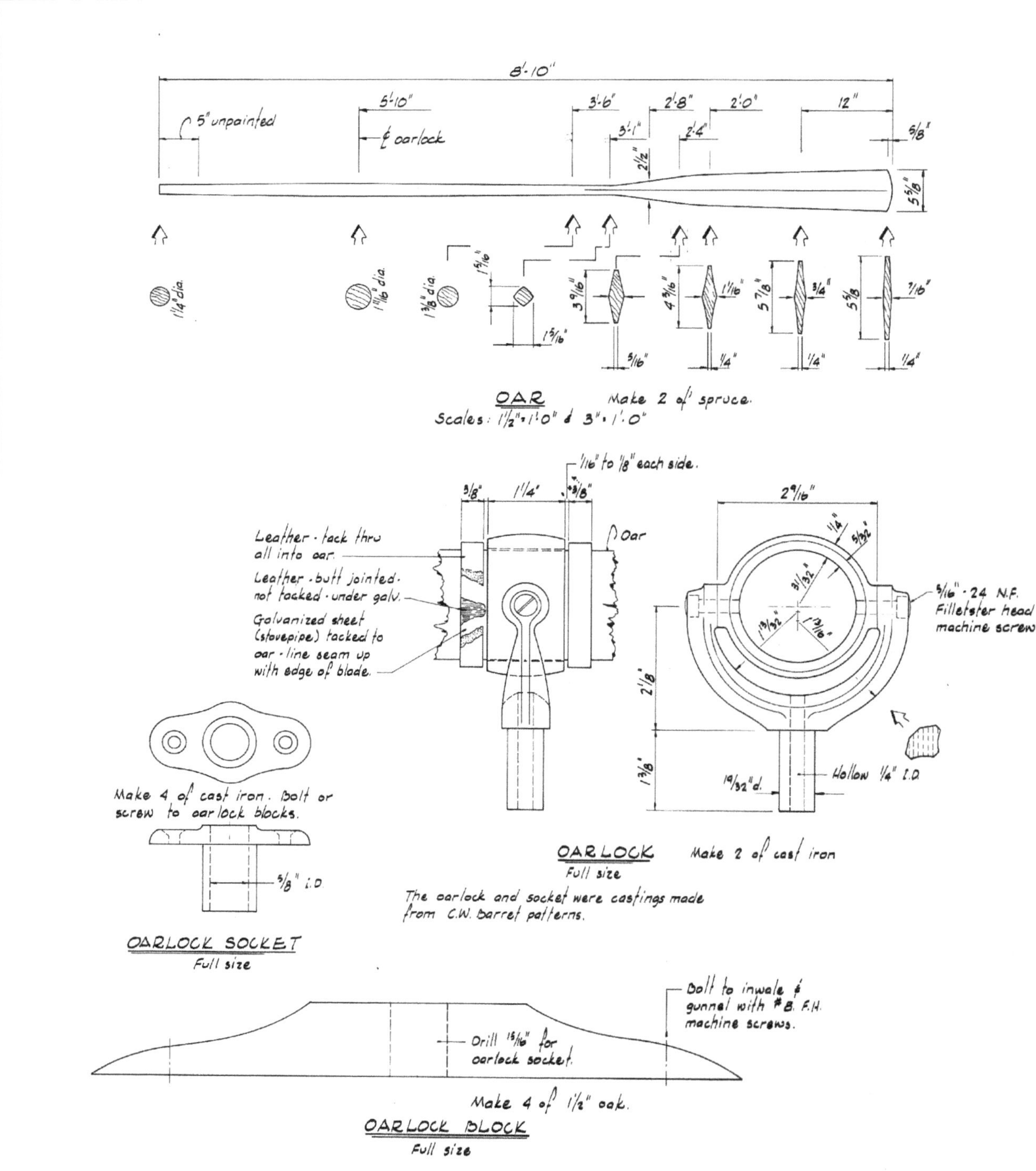

8'-10"
5'-10"
5" unpainted
¢ oarlock
3'-6"
2'-8"
2'-0"
12"
3'-1"
2'-4"
2½"
5/8"
5 5/8"
1¼" dia.
1 3/16" dia.
1 3/8" dia.
1 5/16"
3 9/16"
4 9/16"
1 1/16"
5 7/8"
3/4"
5 5/8"
7/16"
1 5/16"
5/16"
1/4"
1/4"
1/4"
OAR
Make 2 of spruce.
Scales: 1½"=1'-0" & 3"=1'-0"
1/16" to 1/8" each side.
3/8"
1¼"
3/8"
Oar
2 9/16"
1/4"
5/32"
3/32"
13/32"
3/16"
Leather · tack thru all into oar.
Leather · butt jointed. not tacked · under galv.
Galvanized sheet (stovepipe) tacked to oar · line seam up with edge of blade.
3/16" · 24 N.F. Filletster head machine screw.
2 1/8"
1 3/8"
19/32" d.
Hollow 1/4" I.D.
OARLOCK
Make 2 of cast iron
Full size
The oarlock and socket were castings made from C.W. barret patterns.
Make 4 of cast iron. Bolt or screw to oarlock blocks.
5/8" I.D.
OARLOCK SOCKET
Full size
Bolt to inwale & gunnel with #8 F.H. machine screws.
Drill 13/16" for oarlock socket.
Make 4 of 1½" oak.
OARLOCK BLOCK
Full size

PLANK WIDTHS · inside face

Plank	Post**	2	4	6	8	10	12	14	16*	Tran.
K	2¼	1¾	2⅜	2 13/16	3	3⅛	3⅜	2 13/16	2 9/16	1½
J	1⅞	2	2¼	2½	2⅝	2⅞	3⅛	2½	2 5/16	1¾
I	1⅞	2	2½	2¾	3	3	2 13/16	2½	2⅛	2⅝
H	2 1/16	2 1/16	2¼	2½	2¾	2¾	2 9/16	2 5/16	2⅛	2
G	2	2	2½	2⅞	3	2⅞	2 13/16	2 9/16	2½	2⅜
F	2¼	2 3/16	2½	2⅞	3⅛	3	2 13/16	2 13/16	2 9/16	2 3/16
E	2 5/16	2 5/16	2½	2¾	3	3	2¾	2 9/16	2⅜	2½
D	2 13/16	2¼	2 7/16	3	3	2⅞	2⅝	2⅜	2⅜	2 3/16
C	2 13/16	2 3/16	2¾	2⅞	2⅞	2¾	2¾	2⅜	2¼	2½
B	2 13/16	2 7/16	2⅝	2¾	2¾	2⅞	2 9/16	2⅜	2 3/16	2 1/16
Garb.	·	2 13/16	3	3⅛	3 3/16	3	3	3⅛	2⅞	3½

These plank widths are to be used only as a guide in lining off
on the moulds. Finished widths will vary.
The planks are named as Herb Ellis does.
The widths given do not include the 7/8" lap.
The garboard width is from the bearding line.
At the backbone/post joint the top of the garboard is about 3½"
above the bottom of the backbone.
* These widths actually taken 15·2" from F.P.
** Measured along the forward face of the post (rabbet line).

LAP PROJECTION

Plank	2	4	6	8	10	12	14
J	3/32	⅛	⅛	⅛	5/32	⅛	⅛
L	⅛	1/16	1/16	1/16	3/32	⅛	1/16
H	5/32	3/32	1/16	1/16	3/32	1/16	1/16
G	⅛	1/16	3/32	1/16	1/16	1/16	1/16
F	⅛	3/32	3/32	3/32	1/16	3/32	1/16
E	⅛	⅛	3/32	1/16	1/16	⅛	⅛
D	5/32	7/32	3/16	3/16	¼	¼	3/16
C	3/16	¼	7/32	7/32	¼	7/32	3/16
B	3/16	¼	7/32	7/32	¼	3/16	¼
Garb.	3/16	3/16	3/16	3/16	¼	¼	¼

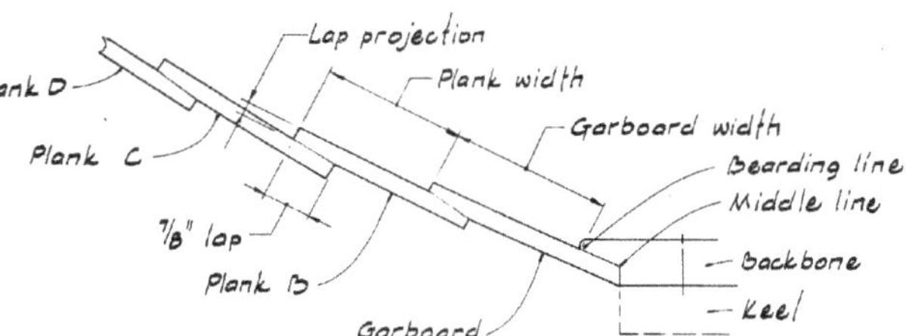

Where the corner of the garboard
is below the backbone, plane it off.

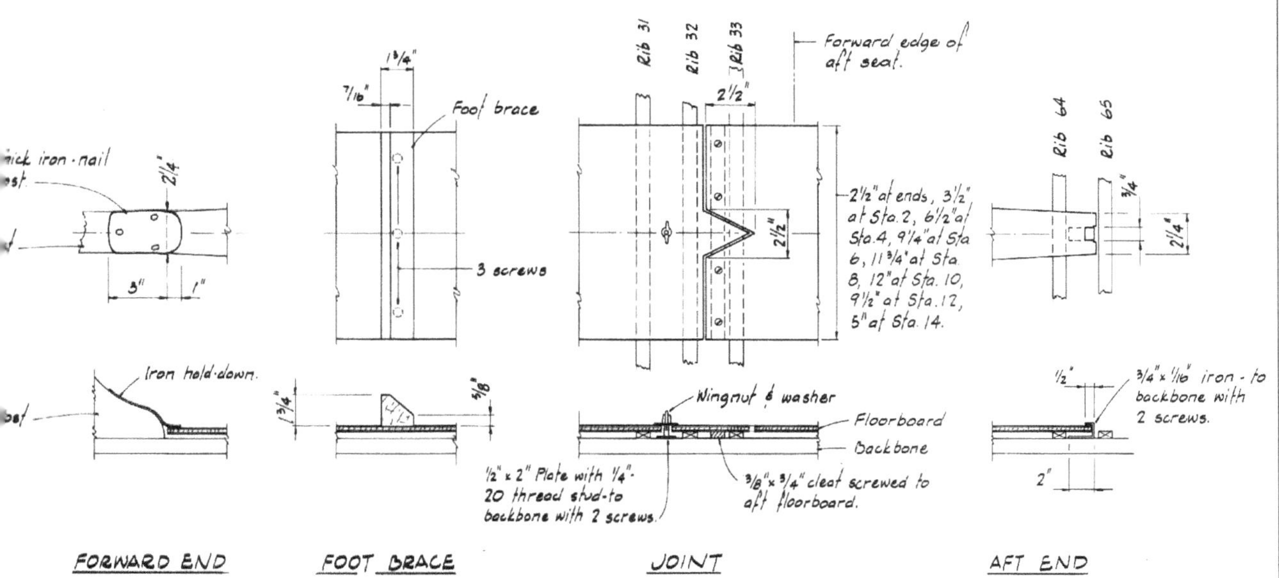

Bibliography

Abbott, Edward. "The Androscoggin Lakes." *Harper's New Monthly Magazine*. June 1877.

Allerton, R. G. *An Account of a Trip of the Oquossoc Angling Association to Northern Maine in June 1869*. New York: Perris & Browne, 1869.

Barker, F. C. *Lake and Forest As I Have Known Them*. Boston: Lothrop, Lee & Shepard Co., 1903.

Barrett, Edwin C. *The History of the Rangeley Boat*. Typescript. The Apprenticeshop of Atlantic Challenge, Rockland, ME.

Bates, Joseph D., Jr. *Streamer Fly Tying and Fishing*. Harrisburg: Stackpole Co., 1966.

Brackett, George Edward. "Notes on Building a Rangeley Boat with Herb Ellis." September/October 1986. The Apprenticeshop of Atlantic Challenge, Rockland, ME.

Bray, Maynard. *Watercraft*. Mystic, CT: Mystic Seaport Museum, 1979.

Cole, Stephen A. Interview with Walter Davenport, Rangeley, Maine. Summer 1988. The Maine Folklife Center, University of Maine.

______. Interview with Harland Kidder, Oquossoc, Maine. Summer 1988. The Maine Folklife Center, University of Maine.

______. Interview with Edwin C. Barrett, Orrington, Maine. Summer 1988. The Maine Folklife Center, University of Maine.

______. "Maine Sporting Camps." Maine Historic Preservation Commission, Augusta, ME, 1989.

Dextor, Almon. *And The Wilderness Blossomed*. Philadelphia: H. W. Fisher & Co., 1901. (Almon Dexter is the pen name of Frederick Dixon.)

Durant, Kenneth and Helen. *The Adirondack Guide-Boat*. Camden, ME: Adirondack Museum and International Marine Publishing Co., 1980.

Ellis, Edward. *A Chronological History of the Rangeley Lakes Region*. Farmington, ME: Rangeley Lakes Historical Society, 1983.

Everhart, W. Harry. *Fishes of Maine*. Augusta: The Maine Department of Inland Fisheries and Wildlife, 1976.

Farrar, C. A. J. *Farrar's Illustrated Guide-Book to the Androscoggin Lakes.* Boston: Lee & Shepard, 1883.

______. *From Lake to Lake.* Jamaica Plain: Jamaica Publishing, 1887.

Forest & Stream. A Weekly Journal of the Rod and Gun. New York.

Foster, Elizabeth. *The Islanders.* Boston: Houghton-Mifflin, 1946.

Gardner, John. *Building Classic Small Craft.* Camden, ME: International Marine Publishing Co., 1977.

______. "Rangeley Boat." *National Fisherman.* January–April 1968, September–October, 1969.

______. "The Rangeley Boat—A Background Sketch." *The Log of Mystic Seaport.* Winter 1973.

Getchell, D. R., Sr. "Fishing with a Rangeley." *Small Boat Journal.* October–November 1986.

Golder, Arthur L. "The Rangeley Lakes." *New England Magazine,* March–August 1900.

Hoar, Jay S. *Small-Town Motion Pictures and Other Sketches of Franklin County, Maine.* Farmington, ME: Knowlton & Hoar, 1969.

Keats, John. *The Skiff and the River.* Nantucket: The Herrick Collection, 1988.

Kendall, William Converse. "The Rangeley Lakes, Maine; with Special Reference to the Habits of the Fishes, Fish Culture and Angling." *Bulletin of the United States Bureau of Fisheries,* Vol. 35 (1915–16).

Lambert, Samuel W., Jr. *The Oquossoc Angling Association 1870–1970.* Privately printed.

Maine. American Guide Series. Boston: Houghton-Mifflin, 1937.

Maine State Year-Book and Legislative Manual. Portland, ME: Hoyt, Fogg & Breed.

Manley, Atwood. *Rushton and His Times in American Canoeing.* Syracuse, NY: Syracuse University Press, 1968.

Mayer, Alfred M., ed. *Sport with Gun and Rod in American Woods and Waters.* New York: Century Co., 1883.

McGuire, Paul. "The Rangeley Tradition." *Wooden Boat,* No. 39, Spring 1981.

Oquossoc Angling Association Account Book, 1869. Oquossoc, Maine.

Rangeley Lakes (newspaper), 1895–97. Rangeley Historical Society, Rangeley, ME.

Rangeley Lakes Region Cultural Inventory, 1998. Maine Arts Commission.

Taylor, David. Interview with Rangeley boatbuilder, Herbert N. Ellis. Winter 1984. Accession 1722, The Maine Folklife Center, University of Maine.

Tidd, Marshall. "Up the Magalloway River in 1861," ed. L. Felix Ranlett and Benton L. Hatch. *Appalachia,* December 1957, June 1958.

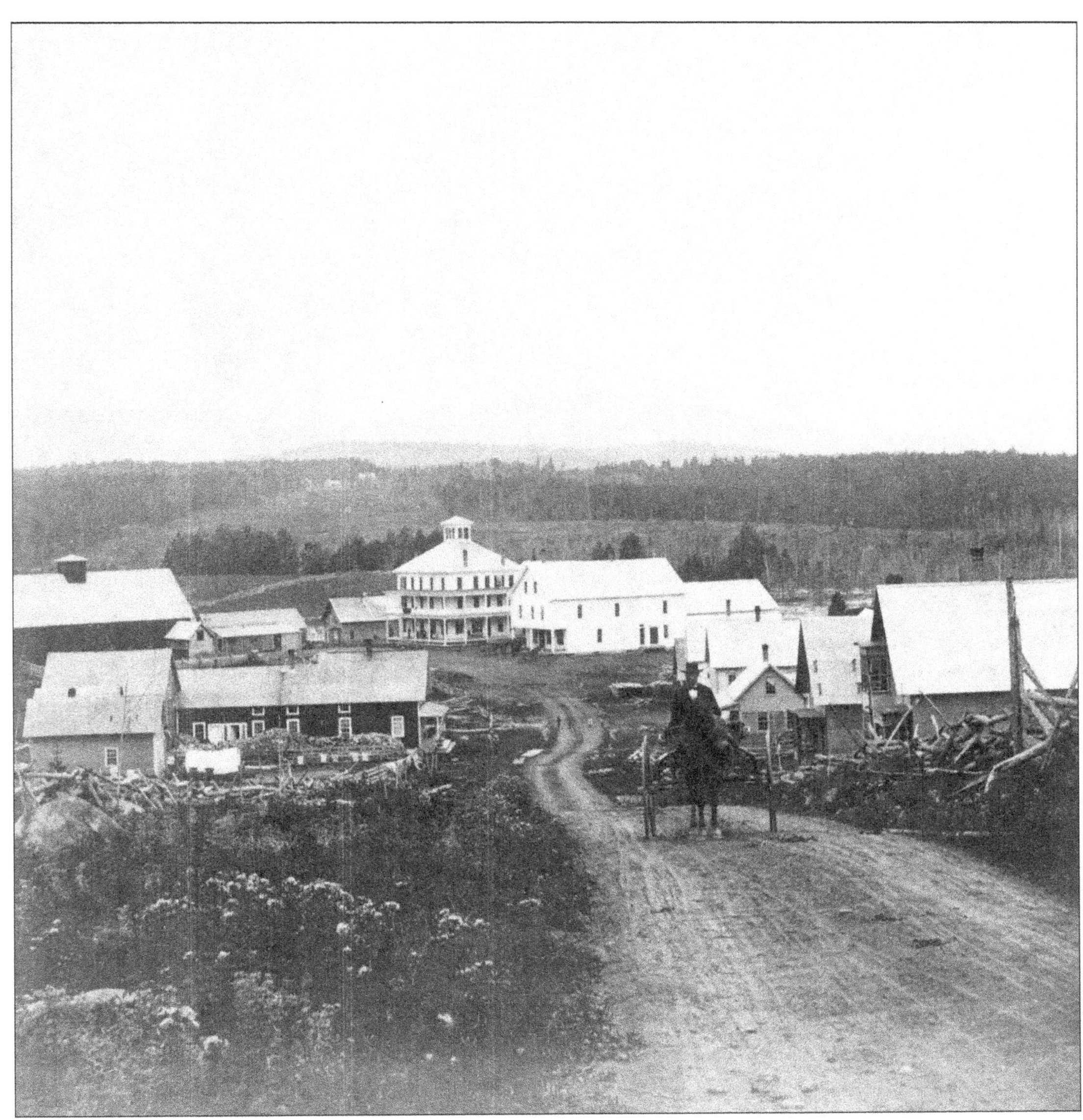

Nineteenth-century Rangeley "City" *(Maine Historic Preservation Commission)*

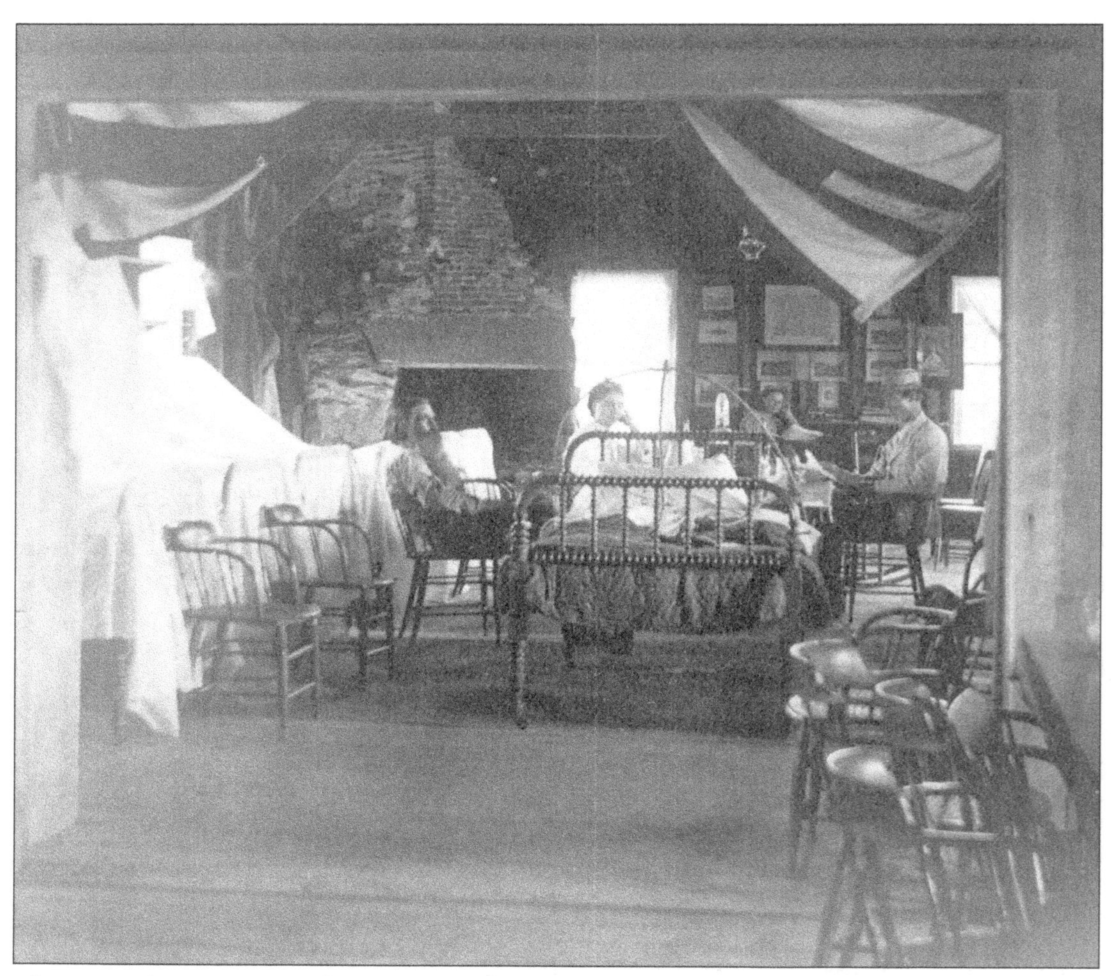

The atmospheric interior of the Oquossoc Angling Association
(Maine Historic Preservation Commission)

Acknowledgements

The Rangeley Lakes and the Rangeley boat have many lovers. I am grateful to the following, both living and departed, for their time and interest in this research: Mr. Walter C. "Skeet" Davenport; Margaret S. Davenport, former director, Rangeley Public Library; Oscar Riddle, former president, Rangeley Historical Society; Joanne Koob, proprietor of Oquossoc's Own B&B; Herbert Ellis; Harland Kidder, former superintendent of the Oquossoc Angling Association; Mrs. Leath Cameron, Art Emory, Grant's Kennebago Camps; Sonny Gibson, proprietor of North Camps; Dick Frost, former proprietor of the Rangeley Region Sport Shop; Edwin C. Barrett; Paul Reagan, Shaw & Tenney; Ben Fuller, Penobscot Marine Museum; John Gardiner, Mystic Seaport Museum; Paul McGuire, Bud Brackett, Fortin Powell, Mr. and Mrs. Montgomery, Phillips Historical Society; Don Palmer and Shirley Adams, the Rangeley Historical Society; Pamela Dean, Maine Folklife Center; Lance Lee and Steve McAllister, the former Rockport Apprenticeshop; Warren Kaericher and Kevin Carney, the Apprenticeshop of Atlantic Challenge; Earle G. Shettleworth, Jr., and Kirk Mohney, Maine Historic Preservation Commission; and Stephen W. Brooke, formerly of the Maine State Museum. A special expression of gratitude goes to Bruce Bauman, whose affection for the Rangeley and its region made this book possible. Finally, eternal thanks to Jennifer Bunting, Tilbury House, Publishers, who saw a book in this manuscript, and to Lindy Gifford, book designer, best friend and spouse, all my love.

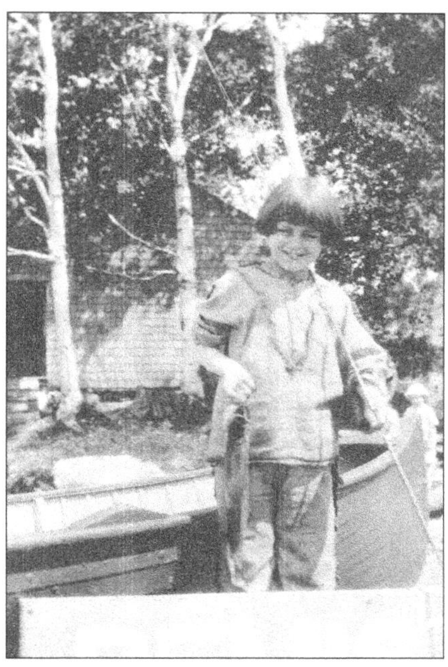

At Bemis *(Rangeley Historical Society)*

A Note on Photographs

Historic photographs used in this book come primarily from two sources, the Rangeley Historical Society and the Maine Historic Preservation Commission. Most from the commission reside in a collection of late-nineteenth-century stereo views representing the following photographers, among others: Frank A. Morrill, New Sharon, ME; G. H. Nickerson, Provincetown, MA; C. A. J. Farrar, Jamaica Plain, MA; Harry P. Dill, Phillips, ME; and E. F. Smith, Boston, MA. The commission also has a large collection of photographs by Edwin R. Starbird, who was based in Farmington, Maine, in the late nineteenth century. Starbird is best known his Woods of Maine series: 600 photos taken around Rangeley, Moosehead Lake, and the West Branch of the Penobscot River. Some of the photographs of Rangeley builder Herb Ellis and his shop were taken by Michael Chaney in 1981 as part of the Maine Folklife Survey (1981–83) and are housed at the Maine Folklife Center, University of Maine.